Grow your Groceries

Grow your Groceries

40 Hacks for Growing Plants from Supermarket Foods

Simon Akeroyd

1.5 million followers on **TikTok** and **Instagram**

Contents

Introduction 6

how to get Growing 8

How plants grow—and how
   we can help 10
Raising new plants for free 12
Grower's starter kit 14
Feeding plants from waste 16

from the Veggie Aisle 18

Green onions from
   the roots 20
Lettuce from the stalk 24
Tomatoes from slices
   of tomato 28
Avocado from a seed 32
Celery from the base 36
Potatoes from tubers 40
Sweet potatoes from slips 44
Corn from
   corn on the cob 48
Leeks from the base 52
Salad leaves from carrot tops 56
Red cabbage from the base 60
Squash and pumpkins
   from seed 64

from the Herb Aisle
68

Mint from cuttings 70
Chile or bell pepper
   from seed 74
Garlic from a clove 78
Lemongrass from a stem 82
Ginger from roots 86
Cilantro by dividing a pot 90

from the Dry Goods Aisle
148

Chamomile from a teabag 150
Lentils from dried seed 154
Peas from dried seed 158
Mustard from the
   whole spice 162
Chia from seed 166
Chickpeas from dried seed 170
Sunflowers from seed 174
Hazelnuts from nuts 178
Quinoa microgreens
   from seed 182

from the Fruit Aisle
94

Strawberries from fresh seed 96
Lemon tree from a pip 100
Plums from a stone 104
Apples from a pip 108
Raspberries from
   fresh seed 112
Pineapple from a cutting 116
Blueberries from seed 120
Passion fruits from seed 124
Mango from a stone 128
Kiwis from seed 132
Melon from seed 136
Lychees from stones 140
Pomegranates from seed 144

Glossary 186
Index 188
Acknowledgments 191
About the author 192

Introduction

Growing your own food is so rewarding. Not only does it save money, but the excitement of seeing your seedlings germinate, or your plant break from dormancy, and produce a flower or fruit is a buzz that never lessens.

Grow Your Groceries features 40 hacks for growing edibles from your supermarket shop, including fruit, veggies, herbs, and even dried seeds and beans, based on many of the videos and reels I've posted on my social media channels. My kitchen windowsills and garden are full of foliage, flowers, and fruits grown from ingredients I have bought in the store.

I've included my top 40 in this book, but there are hundreds of others you can try. Once you've mastered the basic principles, you will be scanning those grocery aisles every time you shop, to see what else you can add to your burgeoning grow-your-own garden. Just a few purchases at the supermarket should set you up with an almost endless supply of seeds, cuttings, and plants, making you as self-sufficient as possible. You could find yourself needing to rely far less on shopping from the big supermarket chains, thereby saving money.

With some fruit and veggies, such as mangoes, pineapples, and avocados, you might not get to harvest any fruit—they can take years to become productive. Some might be sterile or need other pollinators to produce fruit. But it doesn't matter—you will still end up with a beautiful houseplant.

Recycling is at the heart of everything I do when it comes to gardening. Very often it is the part of the ingredient that would usually be thrown away which

is used for growing, for example seeds from a butternut squash or the root base from an onion, leek, or cabbage—meaning you get to eat your shopping and grow it too.

It is not just supermarket produce that can be recycled; it's the containers it is grown in. Everything in this book can be grown in recycled plastic containers or toilet paper tubes, not to mention old plastic flowerpots you already have at home. There is no need to buy expensive terra-cotta or porcelain pots, or pricey potting mix or plant food. It doesn't have to look pretty: the objective is to grow food sustainably and enjoy it. The key ethos behind every hack in this book is sustainability. See the following pages for advice on recycling containers, composting, and getting started.

One final warning, growing plants from the supermarket is addictive. Never again will you look at an apple, onion, or corn on the cob at the store and fail to wonder about the possibilities for growing it. It's fun, exciting, good for your health, and offers a fascinating insight into the food that we so readily accept from the supermarket.

Happy grocery shopping, happy growing, and happy eating.

How plants grow
and how we can help

Many of the ingredients you buy from the fresh fruit and veggie aisles at the supermarket or grocery store are bursting with potential energy, ready to grow and reward you with delicious, edible crops if you provide them with the right conditions.

In fact, it's not just the fresh food aisles. If you look carefully in the frozen and dried food sections there are seeds, beans, frozen fruits, and more, much of which can be propagated from and grown as plants to supply you with more food. All you need to do is bring your groceries home and give them the right environment and some TLC to enable them to grow.

Let there be light
(and some water and warmth)

Plants have three basic requirements in order for them to start growing. These are warmth, light, and water. If you can provide plants (including many of your groceries) with these three conditions, they should thrive.

The good news is that you don't need a garden to grow many of the plants I've included in this book. In many cases all you need is access to a faucet and a sunny windowsill. Some of the larger plants will need larger pots to allow them to get bigger, but that is basically it.

Although most plants require light to grow, be aware that direct sunlight through the window can scorch the foliage of plants. This is particularly the case with young seedlings. Very often a plant requires a "light" room, but not necessarily in direct contact with the sun's rays, which are intensified through glass.

Water seeds after sowing, and after they germinate and start to develop. The amount you water will vary depending on the plant and the temperature. A good way to gauge whether a plant needs watering is to push your finger into the potting mix. If the soil feels dusty and doesn't stick to your finger, then the plant will benefit from a quick drink.

Peat-free potting mix

Throughout this book I have recommended using peat-free, general-purpose potting mix. Commercial manufacturers always used to add peat. This naturally occurring material has taken thousands of years to form, and it provides an essential habitat for rare flora and fauna. Extraction from fragile, finite, and ever-diminishing areas of peat is of major ecological and environmental concern. So when you are buying, look for "peat-free" potting mix. The bags are usually clearly labeled. This means that no peat has been added to the mix.

Of course, the best type of potting media you can use is homemade. Not only does it cost you nothing, but it is a good way of recycling kitchen and garden waste. You don't need much room to make compost. I have a small, rotating dual compost system on a stand. I fill one, and let the other decompose into compost.

The key to good compost is getting a 60:40 ratio of green to brown material. Greens can include any of the following: grass clippings, kitchen scraps (from vegetables and fruit, not dairy or meat), and herbaceous plant materials. Browns include cardboard (remove any plastic tape or labels), fallen leaves (lots in the fall), and wood chips. In order to speed up the rate of decomposition, chop up any material as small as you can.

The other speed accelerator is to turn the compost regularly with a fork (or use a rotating drum like I do—it's much easier). This will add air into the compostable material.

Raising new plants for free

Providing your plants with an environment where they can germinate or grow from cuttings is key to successfully growing your groceries at home. One method of encouraging healthier and better germination rates is to make a small propagator to keep indoors.

Propagators are containers or boxes that provide additional warmth to encourage seeds to germinate and cuttings to produce roots and grow.

They aren't essential for many of the plants you can grow at home. But using one will help your seedlings get off to a faster start, producing earlier crops so you can harvest and enjoy your homegrown food longer.

You can of course buy fancy propagators which have adjustable thermostats, temperatures, and vents—some even come with additional grow lights. However, I've found it is easy to upcycle plastic containers into propagators—the types that usually come with a lid and often contain muffins, cookies, or salads. These otherwise disposable containers will provide a lovely cozy environment to get your groceries growing.

How to make a propagator

- Use a skewer to make small drainage holes in the bottom of your container if it doesn't already have them.
- Place it on a tray to catch any drips from the drainage holes when you water your seeds.
- Fill your propagator with peat-free, general-purpose potting mix and sow your seeds into it.
- Keep it on a sunny windowsill. Your seeds will easily sprout in the additional warmth provided by the propagator.
- Water regularly and check the soil often. It may dry out more quickly than usual in the propagator.
- On hot days, pop the lid open so seedlings don't overheat or get scorched if it is in direct sun.

In winter, additional warmth can be harnessed by positioning your propagator on shelves or a table near a heating vent. Or you can place your propagator on a table on an electric blanket covered with sheets of foil.

How to make a mini greenhouse

Once you get the gardening bug, you will quickly find that all your available space indoors gets filled up with pots and trays of seedlings. Many of us would love additional growing space in a greenhouse or conservatory, but we don't have room.

One handy solution is to repurpose clear plastic storage bins. Put them outside on your patio, balcony, or porch and use them as mini-greenhouses for your extra seedlings. Just remember to place a brick or some weight on them in order to prevent the box blowing away one windy night.

Milk carton cloches

To get your seedlings off to an early start you can use plastic milk cartons as cloches. I find the half-gallon are the perfect size. Cut through three sides, halfway down the bottle. The uncut side acts as a hinge, which you can fold back to give your seedlings light and air, or you can keep it closed to provide extra protection and warmth. Use a skewer to carefully make a few drainage holes in the bottom. You can then fill this section with peat-free potting mix.

Grower's starter kit

You don't need a large budget when you start gardening. Many of the items I use are things I've already got around the house. In fact the majority is surplus packaging, containers, and boxes I've been given when shopping. Instead of putting the materials in the recycling bin, I repurpose them in the garden.

Plant labels

I like to cut plastic milk cartons into ½ x 4-in (1 x 10-cm) strips to use as plant labels. Use a pencil to write plant names on the labels. Pencil doesn't wash off in the rain or during watering, unlike some pens, and you can reuse them the following year simply by rubbing off the writing with an eraser.

Plastic bottle watering cans

Plastic milk cartons can also be used as free watering cans. Use a skewer or a penknife to punch lots of small holes in the lid, making a sprinkler. Then fill the carton with water and use it as a watering can. You can adjust the spray simply by squeezing the carton.

Cardboard seed pots

You don't just have to limit yourself to plastic cartons either. I use the 2-pint (1.4 liter) cardboard ones as well. They often contain plant milk. Lie the carton on its side and make drainage holes with a skewer on the bottom. Then use scissors or a knife to cut away the top. Fill with potting mix and you have a simple container for growing seedlings—ideal for growing herbs on the kitchen windowsill when it is too cold outside.

Toilet paper tube seedling trainers

You don't need to buy expensive terra-cotta or "forever" plastic pots to grow your seedlings in. The cardboard tubes from toilet paper are perfect for the purpose. You can fold the bottoms of the tubes inward so the soil doesn't fall out, but I find it isn't necessary if you simply hold your hand underneath when moving them. Once the roots have grown down it holds the soil together. Not folding the bottom makes it easier for the roots to grow downward too.

When it comes to planting, simply plant the entire thing; you don't need to remove the cardboard. Do make sure all the cardboard is covered with soil or compost. Otherwise, the exposed section above the ground can act like a wick and draw out any moisture in the soil, resulting in plants drying out.

Pallet raised beds

Unless you bulk shop, you are unlikely to receive wooden pallets. But you can often find them for free around the back of stores, supermarkets, and warehouses. Ask the owners; they are often happy for you to take pallets away. Make sure they are heat-treated pallets, not chemically treated ones, because you don't want anything unpleasant leaching into your soil. Look for an HT (heat treated) logo.

> ### Raised bed benefits
> Raised beds are useful if you have poor soil in the garden. You can fill a raised bed with lovely compost that your plants will enjoy growing in. They also help improve drainage, and having a bed a bit higher can save on back-breaking bending over too. They often warm up faster because they are off the ground, meaning you might get crops earlier, and they are less susceptible to frost, because warm air rises.

Pallets can be used to create simple raised beds. Take the top slats off and nail them around the sides to retain the soil. Then simply fill the pallet with potting mix and you have a ready-made, free raised bed. I've also made seed trays, potting benches, and baskets from free pallets.

Drip trays

Many containers, such as ice cream cartons, margarine tubs, and plastic yogurt lids can be used as trays to sit under your plant pots. The trays will catch drips or runoff from the bottom of your pots after they have been watered.

Feeding plants from waste

There is no need to spend money on plant food or fertilizer. It is in abundance all around you. If you live near a park or in the country, you can pick and bring home plants to make into free plant food. Always check with the landowner if necessary, and never take lots of plants—leave plenty for wildlife. You only need a small quantity to make plant food.

My favorite plant to forage for is stinging nettles. The food made from this is high in nitrogen, which will help your plants grow and produce lovely green leaves and stems. It is best to pick stinging nettles in the spring when they are making plenty of growth. You will need to arm yourself with a bag, pruners or scissors, and a pair of gloves. It's also best to wear long sleeves, to avoid getting stung on the arms.

Pick lengths of young stinging nettle stalks, including both stems and leaves. Bring them home and chop them in smaller pieces into a bucket. Once the bucket is full of nettles, top it up with water. Leave outside for a few weeks for the leaves to decompose and rot down in the water.

A warning: this liquid really stinks, so don't keep the bucket by your front door or you won't have many visitors! Once the liquid is a dark brown color, you can decant it into repurposed plastic milk jugs. You can then add this concentrated liquid to your watering can once a week or fortnight. Dilute one part nettle food to ten parts water. Water around the root area of the plant from when it comes into growth until it is flowering or fruiting.

On a much smaller scale, you can collect fruit and vegetable scraps and leave them in water in a sealed container such as an ice cream tub for a few days to decompose, and then dilute it with water and feed your plants with it. I often do this with chopped banana skins, which are high in potassium and one of the key nutrients that plants require to produce flowers, fruit, and seed.

You can also occasionally dilute coffee granules and feed them to your plants. However, coffee is slightly acidic, so only do it occasionally, except for plants that like acidic conditions, such as cranberries and blueberries.

Collect water

There has certainly been an increase in the amount of rain falling from our skies in the last few years. Collect it, because it doesn't cost you anything, and furthermore it is (hopefully) free from any additional chemicals—unlike tap water, which will have been treated. Rain barrels can be attached to downpipes from the gutter to collect rainwater that lands on the roof.

On a much smaller scale, you can leave old containers, buckets, or even mugs outside when it rains and use this to water the seedlings on your windowsills.

Save your teabags and coffee filters

You can dry out used teabags and coffee filters and use them instead of bits of broken pots to place over drainage holes in your pots to stop the potting mix from washing out when you water them.

Collect leaves for leaf mold

Leaf mold is made from decomposed leaves and is great for adding to and bulking up your compost. It's another free resource, courtesy of Mother Nature, and there are millions of leaves dropping from the trees in the fall. Gather them up with a rake and put them into trash bags. Poke a few holes in the bottom of the bags and leave them outside for a year or so. When you come back to them you will have gorgeous leaf mold to use in your potting mixes.

from the Veggie Aisle

Hack.01
Green onions
from the roots

Don't discard the ends of your green onions; those roots will grow again for a bonus harvest of chive-like stems. It's almost impossible not to get green onions sprouting. So, if you don't have confidence in your green fingers just yet, this is the perfect beginner plant.

This technique seems to work at any time of year, with plants taking up next to no room on the windowsill. You will be harvesting the fresh, green onion stems produced from the root plate, which provide a delicious, mild onion flavor in salads and savory smoothies. Plants are unlikely to produce mini onion bulbs. However, if you leave them long enough, they will eventually go to seed. And if you collect these seeds and sow in spring, the new plants will produce small bulbs.

You will need

- Green onion bunch
- Knife
- Plant tray
- Peat-free, general-purpose potting mix
- Toilet paper tubes (optional)

Know your plants

Also known as salad onions, spring onions, or scallions, the green onions you're most likely to find in the supermarket are simply normal onions—technical name *Allium cepa*—harvested at an early stage of development, before they've had a chance to form the distinctive bulbs that we peel and dice. If you're lucky, you may also find Welsh onions (*Allium fistulosum*), which naturally form bunches rather than bulbs, and some consider the "true" spring onion.

how to grow

01 Take your green onion bunch and chop them off at the base, retaining about ¾ in (2 cm) of the white stem. If you look carefully, you will almost certainly see small roots at its base.

02 Fill a plant tray with peat-free, general-purpose potting mix. Green onions are shallow rooting, so they only require soil to a depth of about 2 in (5 cm). I often recycle a supermarket container, adding drainage holes.

03 Push the green onions into the soil by ½ in (1 cm) so just a stub appears above the surface, spacing each one 1¼ in (3 cm) apart. If you have a dibble, you can use it to make holes.

04 If you'd like tender white stems, place toilet paper tubes around each base. These protect the stems from light, causing them to remain white, a technique known as "blanching."

FROM THE VEGGIE AISLE

05 Keep on a sunny windowsill and water twice a week. Within a week you will start to see growth.

06 After about 20 days you will have fresh stems to harvest: simply snip them off with scissors. They will regrow after cutting. I've had the same green onion plants on my windowsill for more than a year now.

Freezer supply

If you want to store your green onion stems for another time, chop the stems into ½-in (1-cm) sections, add them to a freezer-proof bottle, and place the bottle in the freezer. Amazingly, the chopped stems don't seem to stick together. There's no need to defrost, simply sprinkle directly over your dish. You can apply the same technique with chive stems. I often do this at the end of the season, for an almost endless supply of oniony stems in the freezer.

Let it thrive

Green onions can be planted outside in the garden too. Keep harvesting and they should continue to regrow throughout the summer. Keep the area free of weeds to stop them from competing for nutrients. I find planting green onions (any onions for that matter) next to my carrots helps keep away carrot fly (an annoying root pest that can destroy your carrots).

Hack.02
Lettuce
from the stalk

No salad is complete without some sort of lettuce leaves in my opinion, and whether you prefer iceberg, butter, or Romaine lettuce, you are spoiled for choice. Luckily for us, they are all easy to regrow from their bases—meaning you get two lettuces for the price of one.

Lettuces are fast growing, so they are a satisfying crop to grow for quick results. Always look for the freshest lettuce in the supermarket. You need one with the remains of a base or stalk (where the plant was cut away from the stem by the grower). Your lettuce will regrow delicious new, upright leaves, but are unlikely to re-create the original ball shape if you have chosen one of the tight-headed types such as an iceberg. Lettuces can be grown in a pot inside or outside, or you can grow them directly in the soil in your garden. They prefer full sunshine, but will tolerate some shade.

You will need

+ One lettuce
+ Sharp knife
+ Small bowl of water
+ Peat-free, general-purpose potting mix
+ Plant pots

how to grow

01 Cut the base of the lettuce away, leaving a 1¼-in (3-cm) stump.

02 Place the base into a bowl of water about ½ in (1 cm) deep. Leave on a sunny windowsill. In about 10 days you will see new leaves emerging from the base of the lettuce.

03 Fill a pot with peat-free potting mix and plant the base of the lettuce at the same depth in the soil as it was in the water.

04 Place either on a sunny patio or keep on your windowsill.

FROM THE VEGGIE AISLE

05 Alternatively, if you have a garden, they can be planted out directly into the soil, spaced 12 in (30 cm) apart.

Let it thrive

Water the lettuce a couple times a week if growing inside, or during dry spells if outside. Keep a lookout for slugs and snails, which love fresh lettuce leaves. Remove them by hand if you see them and place them where they can feed on weed leaves, or by the compost heap. Harvest the young, tender, lettuce leaves as they grow. Use a knife or pruners to cut the leaves near the base. They should continue to regrow a few times after harvesting. Once they seem to have exhausted all their growing energy you can remove the lettuce from its pot and add it to your compost heap.

Hack.03
Tomatoes
from slices of tomato

Nothing tastes better, in my opinion, than a tomato picked fresh from the vine that has been grown in your own soil and warmed by the sun—and they are so easy to grow. Enjoy the taste of the Mediterranean—starting with tomatoes from the store.

Don't be put off if you think supermarket tomatoes don't taste that great; this is usually because they are picked too early and kept in storage for too long. The varieties themselves (of *Solanum lycopersicum*) are delicious. Getting tomato seeds to germinate is practically foolproof using this method. You might think it is a bit like adding tomato toppings to an earthy pizza—and the technique really is as simple as that.

You will need

- Tomato
- Knife
- Old grape container or other vessel
- Peat-free, general-purpose potting mix
- Pots

how to grow

01 Cut a few tomato slices about ⅛ in (3 mm) thick. If you look closely you will see all the tiny seeds contained within each slice.

02 Fill a container with peat-free potting mix to a depth of 1¼ in (3 cm). I used old plastic grape containers. Whatever you use, make sure there are some drainage holes in the bottom.

03 Place a couple of tomato slices on the soil. Cover the slices with more potting mix to a depth of about ¾ in (2 cm).

04 Place on a warm sunny windowsill. Soon you will see lots of seedlings poking up through the soil.

FROM THE VEGGIE AISLE

05 When the tomato plants have formed "true" leaves, they can be transplanted into individual pots. True leaves are different to the first pair of leaves you see when seedlings germinate. You will notice that they look distinct from the first leaves.

06 Keep the tomatoes in a warm, sunny position until they are about 6 in (15 cm) high. Tomatoes are tender, so can be planted outside once the risk of frost is over. You can grow them in a container or two per grow bag; I like to grow them directly in the soil.

Let it thrive

Tomatoes are usually either "cordon" types or bush types. All the tomatoes I have grown from the supermarket have been the former. Each cordon tomato plant will need a stake to support it as it grows. Pinch out the tops when they get to about 6½ ft (2 m) high to encourage a bushier plant below (and more fruit). Tomatoes will be ripe from midsummer onward. If you are lucky enough to have a greenhouse or conservatory you can grow tomatoes there, and you will have a far longer growing season than if they were outside.

Hack.04
Avocado
from a seed

The avocado (*Persea americana*) is easy to grow from its large stone as a houseplant with attractive long, glossy leaves, but it is not so easy to get it to fruit in cooler climates—avocados originate from Central and South America and require a warm climate to bear fruit.

Technically avocados are fruits because their stone (seed) is in the center of their flesh, but they are of course used in savory dishes, and found in the vegetable aisle. If you live in a cool climate, it is unlikely your avocado plant will ever produce fruit. That said, a gardener I know does get avocados from a plant that she sowed a few years ago, but she keeps the plant in a warm conservatory. You will probably have to be content with growing a beautiful houseplant and, if you get fruit, that's a bonus (it could take a number of years to produce fruit too).

You will need

- Avocado
- Knife
- Plastic tub or freezer bag
- 6–7-in (2-liter) pot
- Peat-free, general-purpose potting mix

Grow your avocado seed for beautiful foliage; it will need heat to produce fruit.

how to grow

01 Cut the avocado in half and carefully remove the large stone. Rinse the stone.

02 Put the stone in a freezer bag or plastic tub. Put the bag in a dark cupboard and forget about it for a few weeks.

03 You will with luck discover that the avocado has started to germinate and has sent out a small shoot.

04 Plant the stone in peat-free potting mix about ¾ in (2 cm) deep, with the shoot pointing upward. Water every few days. In a few weeks, the shoot will develop leaves and a single trunk.

FROM THE VEGGIE AISLE

05 Keep your avocado in a warm sunny location indoors and water it occasionally when the soil feels dry to the touch.

06 Feed with a general-purpose liquid food once a week or if the leaves start to turn yellow.

Let it thrive

Avocados in their natural environment (a tropical climate) will grow into fairly large trees. If you are growing yours in a pot indoors as a houseplant, pinch the growing tip out when it reaches 16 in (40 cm) high. This will encourage bushier growth and restrict its size.

Hack.05
Celery
from the base

I love the crunchy stems of fresh celery (*Apium graveolens*). The base, usually thrown away, can easily be regrown at home. The new stems won't be as large as the ones you bought from the supermarket, but your smaller produce will be more succulent and a delicious addition to any salad or soup.

Celery is closely related to parsley and carrots and, like carrots, it is grown for its edible stems. It is best grown in spring so you can enjoy sticks of this refreshing vegetable throughout summer. However, I have also done this in the fall and planted the base outside, where it survived the winter and produced stems the following spring. Its natural habitat is damp, marshy conditions, so make sure you keep your plant well watered once it is established.

You will need

- Head of celery
- Knife
- Large jar or bowl
- Peat-free, general-purpose potting mix
- Pot

how to grow

01 Cut the base away from a whole head of celery.

02 Place the base in a jar or bowl with ½–¾ in (1–2 cm) of water.

03 Leave on a sunny windowsill for a week or two.

04 You will notice stems starting to emerge from the center of the base. You may also see new roots emerging from the base in the water.

05 Plant the base in a pot filled with peat-free potting mix and it will continue to grow new stems. Harvest the stems when they reach 4–6 in (10–15 cm) high. They should continue to produce stems throughout summer if you keep picking and keep the plant watered every couple of days.

06 You can also plant the base outside, directly into the soil or a raised bed. If you have more than one, leave 12 in (30 cm) between each plant.

Blanching celery

If you want the white stems you see in stores, then you will need to blanch them. To do this, use a rake to pile ("hill up") the soil around the stems as they start to grow. Blocking out the sunlight should cause the stems to stay pale.

Hack.06
Potatoes
from tubers

This is possibly the most popular vegetable on the planet—and growing a potato (*Solanum tuberosum*) from the supermarket couldn't be easier. If you leave your potato in a bag in the back of a cabinet, you will find that it has started to send out shoots. Although this is not great for cooking, it is perfect for planting outside to produce more potatoes.

Potatoes are the swollen root system of the plant, known as tubers, so your harvest will be underground—and you will need some outside space for this hack. To save on the back-breaking work of digging in the soil, I like to grow my potatoes in trash bags or old potting mix bags, because they take up hardly any space on my patio. You can of course plant them in containers and raised beds or buy special potato-growing bags.

You will need

+ Potatoes
+ Trash bag or old potting mix bag
+ Peat-free, general-purpose potting mix

If you're organized, you can sow every couple of weeks for a long cropping season.

how to grow

01 Leave the potatoes in a dark cupboard until they start to send out shoots; an old egg carton is a handy way to store them.

02 Add 4 in (10 cm) peat-free potting mix to a large pot, trash bag, or potting mix bag. If using a bag, poke drainage holes in the bottom and roll down the sides about halfway.

03 Put three potatoes on the soil, followed by another 4 in (10 cm) potting mix. Place somewhere sunny outdoors and water weekly if it hasn't rained for a few days.

Hilling up

As shoots and leaves emerge, you will need to "hill up" the potatoes by adding more soil to just below the tops of the foliage. This will stop the new potatoes from turning green, and will make more space for tubers to grow. If you're growing in a bag, roll up the sides as you add more soil. Stop hilling up once the soil is at the top of the pot or bag.

04 Hill up the potatoes once shoots emerge (see box). When flowers appear they can be harvested for early new potatoes; simply root around to find them, or dump out the contents of the bag. If you leave them longer they will grow larger, as you choose. The tubers will stop developing once the foliage has died down.

05 If growing in a pot, you will need to water the potatoes more often than those in the ground. Your potatoes won't grow as large as those grown in the ground, so harvest when new.

Let it thrive

Potatoes originate from South America, and are hardy and suitable for cooler and often rainier climates. I like to start planting potatoes in mid-spring, because this gives you most of the season for the potatoes to grow. If you have room, you can plant a few potatoes every couple of weeks until midsummer to ensure you have a succession of crops throughout most of the growing season from midsummer to late fall.

Hack.07
Sweet potatoes *from slips*

Just like the humble potato, sweet potatoes are grown for their underground tubers. I love this vegetable (*Ipomoea batatas*) because it is so versatile and a great alternative to traditional potatoes. This isn't a crop you can grow on the windowsill; they will need room outside or in a greenhouse.

Because sweet potatoes originate from warm climates, they will need a long growing season, so get them started in early spring to give them enough time to develop tubers underground for harvesting later in the year. They are grown from "slips," which are basically like softwood cuttings from a plant. Start the process of encouraging the sweet potatoes to develop slips as early as possible in spring to ensure they have time to produce tubers before the fall frosts.

You will need

- Container such as for grapes
- Peat-free, general-purpose potting mix
- Sweet potato
- Knife

Fringe benefit

One bonus of growing sweet potatoes is that their leaves and shoots can also be harvested and cooked, a bit like spinach. However, please do not do this with ordinary white potatoes, because their leaves are poisonous!

Sweet potatoes are pretty large, so they will need growing space outside.

how to grow

01 Fill a container with potting mix and make sure it has drainage holes. Wash your sweet potato. Don't bury it, but nestle it horizontally into the soil so the lowest third is covered.

02 Place the container in a sunny position indoors. After a couple weeks you will see shoots emerging from around the base of the sweet potato. These shoots are your slips.

03 When the slips are 4 in (10 cm) tall they can be carefully removed from the original sweet potato with a knife. Put each slip into a jar of water and continue to grow them on your windowsill, ensuring they are kept warm.

04 Once the risk of late spring frosts is over, plant slips outside in containers or directly into the soil. If you have a greenhouse, plant them there; protection from fall frosts gives maximum chance of producing tubers.

FROM THE VEGGIE AISLE

05 These are trailing plants, so give them plenty of space to sprawl. To save space you can train them up bamboo sticks or a trellis.

06 Sweet potatoes will be ready at the end of summer or early fall. Harvest the tubers in the same way you would traditional potatoes by uprooting the plant and digging around in the soil.

Let it thrive

Sweet potatoes are more tender than their distant cousin, the potato, so they need protection from frosts. In really cold areas you will need to grow them under a cover (such as a fleece or cloche) during spring and early fall; if you live somewhere that rarely gets frosts you can grow them outside without a cover. They will be fine in pots and raised beds and could be grown on balconies, roof gardens, and even a porch if they receive plenty of sunlight.

Hack.08
Corn
from corn on the cob

If you love throwing a fresh cob on the grill in summer, then consider this hack. Corn on the cob appears in stores throughout summer, and drying the kernels is the easiest way to grow your own corn at home. Save a cob each year to dry for seeds to use the following season, and you'll have homegrown corn for ever! And you can use this method for popcorn too.

Leave your supermarket cobs to dry out naturally until the individual kernels (which contain the seeds) turn hard, making them suitable for sowing. Once the kernels are dry and hard they can be removed. I find the easiest way to do this is to hold the cob over a bowl and rub my thumb over them so they ping off and into the container below. You'll end up with hundreds of kernels, but unless you are a farmer you only need to collect a small handful. Pop the dried kernels into an envelope and store in a cool dark place until the following spring.

You will need

+ Dried corn kernels
+ Small pots or toilet paper tubes (optional)
+ Peat-free, general-purpose potting mix

how to grow

01 Sow kernels indoors into individual pots or toilet paper tubes using peat-free potting mix at 1 in (2.5 cm) deep, a few weeks before the risk of frost is over.

02 Transplant outside when the seedlings are 4 in (10 cm) high. Corn should be planted in a grid, not single rows, to allow for better pollination rates.

03 Sow kernels directly into the soil outside in a weed-free area. Plant or sow seeds 16 in (40 cm) apart.

04 Sow two seeds in each "square" of the grid to allow for one failing to germinate. Remove the weaker one later if they both start to grow.

05 Your delicious, succulent corn on the cob should be ready from midsummer onward. Enjoy!

Let it thrive

Sweet corn plants are slightly tender, so if you live in a cool climate you need to wait until the risk of frost is over before planting them outside. Germination takes a few weeks, so it's best to sow about three weeks before the risk of frost is over in your area. Corn can get quite large (6½ ft/2 m high), so you will need a garden with, ideally, enough space for at least nine plants to ensure good pollination. They are wind pollinated and therefore need a bit of a breeze.

Pop your own corn

Popcorn is closely related to sweet corn and it can be grown in almost exactly the same way from the packets of popcorn kernels you buy to pop in a pan at home. The plants will produce very similar-looking cobs to sweet corn, but they are far starchier and wouldn't taste particularly good if you try to eat them! Instead, harvest and dry the kernels and fry them in oil to make popcorn (or use them in a popcorn-making machine). Don't grow both sweet corn and popcorn in the same garden, because they will cross-pollinate. This could mean that you end up with popcorn that won't pop and sweet corn that isn't very sweet!

Hack.09
Leeks
from the base

Supermarket leeks (*Allium porrum*) can be grown from the scraps that you would otherwise throw away. The ones I've grown taste just as good as those I've bought from the store. You can grow your leeks directly in the ground if you have a garden, or you can keep them on a windowsill and harvest them when you are next having a roast.

If you like the idea of growing food at home then leeks are a must, because they are one of the few vegetables tough enough to hang around outside when it gets cold. Leeks are hardy, so they can stay out in the ground during winter until you are ready to harvest them. They require a long growing season, so if you wish to enjoy these vegetables over fall and winter, which is the traditional time to harvest them, you will need to start growing them in spring.

You will need
- Leek
- Knife
- Jar or saucer
- Pot
- Peat-free, general-purpose potting mix
- Dibble or trowel (optional)

how to grow

01 Remove the base of a leek with a knife, leaving a stub of about ¾ in (2 cm). Use the remainder of the vegetable to cook with.

02 Place the leek into ½ in (1 cm) of water in a jar and leave on a sunny windowsill. Top up the water so there is always ½ in (1 cm).

03 Soon your leek will start to produce roots and the stem will shoot upward. When the mini leek has reached 4 in (10 cm) it can be planted into potting mix, at the same depth as it was in the water. Keep in a sunny spot.

Leeks for dried flowers

Leeks are herbaceous perennials, although most gardeners grow them as an annual. If you leave the leek in the ground, it should produce more stems next season. My tip, however, is to allow the leeks to go to flower because they have huge seed heads that you can dry and use indoors as an impressive dried-flower display. You can of course collect seeds from the flower heads and sow more leeks the following year.

FROM THE VEGGIE AISLE

04 If you intend on growing leeks outdoors in the ground then use a large dibble or trowel to create a hole that is about 2 in (5 cm) deep and 1¼ in (3 cm) wide.

05 Place a leek into the hole and use a watering can to gently "wash" the soil back around it. Once established, continue to water regularly.

Let it thrive

The reason for filling the planting hole by "washing" the soil in is that the looseness of the surrounding soil will allow the stem of the leek to thicken easily without too much resistance from firmer soil. In addition, planting the leek a bit deeper will ensure that the emerging stem of the leek is white and succulent, just like the ones you buy in the supermarket. Exposure to sunlight causes the plant to produce greener stems.

Hack.10
Salad leaves
from carrot tops

Carrots are of course grown for their earthy-flavored roots, but leafy green carrot tops are also edible and taste similar to parsley. To grow your own leaves at home, you need to buy carrots in the bunch with their leafy tops still attached.

I've found the best time to get started growing carrot tops for salad leaves is in spring, when they can be planted directly outside if you have room. If you are doing this in the fall, I find it best to plant them indoors in pots and keep them in a sunny position. If they are kept warm, they should overwinter and continue to produce foliage the following spring. It is unlikely that your planted tops will produce another root below ground. That said, some followers on my social media accounts have commented that they have managed to encourage carrot tops to produce further edible roots. So it's worth giving it a try.

You will need

+ Three carrots
+ Knife and scissors
+ Pot with drainage holes
+ Peat-free, general-purpose potting mix

how to grow

01 Cut the tops off the carrots, keeping about ¾ in (2 cm) attached to the leafy foliage. Trim the foliage back to about ½ in (1 cm).

02 Fill a pot or container with peat free, general-purpose potting mix and plant the carrots with their tops just level with the surface. Grow them on your windowsill or outside during summer.

03 In a few days you will see the new foliage starting to emerge from the tops of the carrots.

04 If growing in a pot outside, plant with the tops of the carrots level with the surface, and place the container in full sun.

Let it thrive

You can also plant carrots directly into the ground outside in spring or summer. Carrots require a light soil and might rot if you plant them in a heavy clay soil. If your soil is clay, then it is best to grow them in containers.

05 When the leaves are about 4 in (10 cm) high they can be harvested and used instead of parsley in recipes. The carrots should continue to produce additional flushes of foliage throughout the summer.

Beet it

This method also works with beets—the foliage makes delicate salad leaves. You'll need to buy beets in a bunch. Remove the leaves (use them in cooking), leaving about ½ in (1 cm) of the stalks attached. Slice 1¼ in (3 cm) off the top of the beets. Plant this slice in peat free, general-purpose potting mix with the top just level with the surface. Grow on a windowsill or outside during summer.

Hack.11
Red cabbage
from the base

Red cabbage is one of the most colorful vegetables in the grocery aisle. If your homegrown cabbage looks purple and not red, it is because the color can vary depending on the pH of the soil it grows in. In acidic soil (low pH) the leaves are usually redder and in more neutral and alkaline soils (high pH) the leaves tend to become more purple.

Cabbages are hungry plants, so add plenty of organic matter such as compost or manure to the soil prior to planting outdoors. If you are planting your cabbage indoors in a pot, then feed once or twice a week while it is growing with a balanced, liquid plant food. Toward the end of the season, you can leave the cabbage to flower and set seeds. These seeds can be harvested and sown the following spring, and those seedlings will form into a tight red ball like the original cabbage. Red cabbages are annuals, so you can dig them up (or pull them up if in a container) and add them to your compost heap at the end of the season.

You will need

+ Red cabbage
+ Knife
+ Shallow dish
+ Pot
+ Peat-free, general-purpose potting mix

how to grow

01 Cut off the base of the red cabbage, leaving a stub of about ¾ in (2 cm). Put it in a dish or jar with ½ in (1 cm) of water and place on a sunny windowsill.

02 Change the water every few days; fresh water helps encourage rooting. After one or two weeks you will see new leaves starting to emerge from the base.

03 The cabbage can now be planted in peat-free potting mix and kept on the windowsill.

FROM THE VEGGIE AISLE

04 Alternatively, plant outside, about ½ in (1 cm) below the soil. If you are planting more than one, space the cabbages 12 in (30 cm) apart.

Let it thrive

Keep the plant watered during dry periods. The cabbage won't form into a solid ball like the original plant, but instead will give you leaves, and should continue to produce fresh leaves after harvesting. Keep picking to ensure it keeps supplying you with fresh cabbage leaves throughout summer and into the fall.

Brassica bonus

This same method of propagating a plant from its base can be followed with other members of the cabbage family, or brassicas, including bok choy, Savoy cabbages, kale, and Chinese cabbage. I have managed to root a broccoli stem as well, although that was harder to do. I have also used the same method for growing all types of lettuce from the supermarket.

Hack.12
Squash
and pumpkins from seed

Squash are one of the most popular fall vegetables, and if you collect the seed you can grow your own the following year and never buy one again. Of course, many of us buy a pumpkin every year for carving at Halloween. You can use the same method to grow your own.

Squash are easy to grow from the seeds you can scoop out from the inside. They are annual climbing or trailing plants, so you will need to sow each year in order to produce fruits in the fall. Growing a squash has two parts: in the fall, collecting and storing the seeds for more than half a year, and then in late spring sowing and growing the seeds. If you are growing a pumpkin, you will almost certainly purchase it in the fall because this is when they are available for sale.

You will need

+ Squash or pumpkin
+ Sharp knife
+ Strainer
+ Paper towels
+ 3½-in (9-cm) pots
+ Peat-free, general-purpose potting mix

Grow as microgreens

Squash plants are large and sprawling, so you will need an outdoor space if you want to grow them. However, it is possible to grow the seed as microgreens at any time of the year. Simply sprinkle a few seeds over a seed tray filled with peat-free, general-purpose potting mix and leave on a sunny windowsill. After a couple weeks, they will produce shoots and leaves that can be picked and added to salads.

how to grow

01 Cut the squash in half—or slice off the top if you are carving a pumpkin for Halloween—and scoop out some of the seeds.

02 Place them in a strainer and give rinse them to remove as much of the pulp as possible. Leave them to dry on paper towels for a few hours.

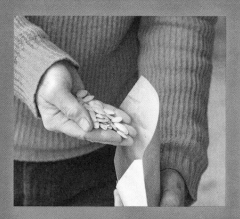

03 Put the dry seeds in an envelope, label them, and store somewhere cool and dry until mid-spring, when they can be sown for planting out in late spring or early summer.

04 Sow one seed per 3½-in (9-cm) pot in peat-free, general-purpose potting mix. Squash and pumpkin seeds are best planted on their edge ¾–1¼ in (2–3 cm) below the surface. Otherwise, water collects on their broad surface and they have a tendency to rot.

05 Once the seedlings are 4 in (10 cm) high they can be planted outside in fertile soil (see box). They take up a lot of room, so if you are planting several give them plenty of space apart from each other.

06 Squash can be trained up trellis, fencing, and walls by tying in the shoots with string. You may need to provide extra support for the swelling fruit if they get very heavy.

Let it thrive

Squash and pumpkins are tender plants, so don't plant out your seedlings until the risk of frost is over. This is usually late spring or early summer. They like a rich and fertile soil; add plenty of organic matter to the area prior to planting. Site individual plants at least 3 ft (1 m) apart, and ideally even farther. They will sprawl on the ground, but you can train them onto a trellis to save space. If you want especially large pumpkins for Halloween, then remove some of the emerging pumpkins and allow just two or three to form on the plant. Give the plants a liquid food every week (a standard tomato food is good) around the root area, starting when the flowers appear through to late summer. As fall approaches, you can place a paving slab under the heavy fruit so it doesn't start to rot on the ground before you are ready to harvest.

from the Herb Aisle

Hack.13
Mint
from cuttings

This traditional garden herb (*Mentha* x *piperita*) is loved for its fresh-tasting leaves and can easily be grown from the sprigs in a supermarket package. Buy the freshest available, with the longest best-before date. While peppermint is ubiquitous, there are many different flavors belonging to the mint family. In my own garden I grow ginger mint, strawberry mint, chocolate mint, and even banana mint.

The great news is that mint is very easy to grow. The not-so-good news is that mint is almost too easy to grow! If you aren't careful, it can take over your entire garden as it sends out its underground rhizomes: vigorous rootlike stems that will grow through the soil and send out new shoots. For this reason, always grow mint in a container or pot and never plant it directly into the ground. Unlike many of the other culinary herbs, which generally prefer dry, arid conditions, mint isn't as picky and is quite happy in moderately moist soil, so don't forget to water the plant regularly during the summer.

You will need

- Fresh mint stems
- Sharp knife
- Glass or jar
- 3½-in (9-cm) pots
- Peat-free, general-purpose potting mix
- Pots of increasing sizes

how to grow

01 Remove the lowest two-thirds of the foliage along the stem. You can do this simply by running the stem between your thumb and index finger. Save the leaves for cooking later.

02 Use a knife to trim the end of the stem, cutting just below a node (bud or bumpy part of the stem).

03 Place these cut stems of mint into a glass or jar, ensuring the bottom ¾ in (2 cm) of the stems are covered with water.

04 Place the jar of water on a sunny windowsill and refresh the water every few days. Roots should appear at the bottom of the stems after two or three weeks.

05 Plant individual stems into 3½-in (9-cm) pots in peat-free, general-purpose potting mix. Keep on your windowsill or on the patio outside.

06 Repot plants into larger containers as they get bigger. After two or three years you can remove the plant from the pot, divide the roots into smaller clumps and plant these in more pots.

Let it thrive

Keep an eye on your mint plants to make sure they haven't spread. I've seen mint plants in pots that have been plunged into borders and yet have still managed to spread out of drainage holes or send shoots over the top of the pot and root in the surrounding soil. When grown in pots, mint can be short-lived, so I suggest taking cuttings regularly and repeating the process above to keep a regular supply of mint.

Hack.14
Chile
or bell pepper from seed

Chiles and their milder relatives, the bell peppers, are easy to germinate from their seeds, and they reliably produce colorful and spicy fruits each year in my garden. Although there are hundreds of amazing chile varieties, only a few types appear in the supermarket. I like to visit farmer's markets and farm stores in summer to find rare or unusual chiles I can propagate from.

Chiles require a fairly long growing season, so seeds should be sown in late winter or early spring. If you have a greenhouse or will be growing them indoors then you can start sowing as early as January to make the most of the extra frost protection under glass. If you plan on growing chiles outside in containers or directly in the ground during summer, then wait until March to sow, because otherwise the plants will be ready too early and they won't be able to face the cold. You can use the same technique for bell peppers. Select red bell peppers because they generally have riper seeds. This isn't always the case with chiles, which come in a range of colors: red doesn't always denote ripeness.

You will need

+ Fresh chile or red bell pepper
+ Sharp knife
+ Paper towels
+ Seed tray
+ Peat-free, general-purpose potting mix
+ Propagator (optional, see p.13)
+ 3½-in (9-cm) pots

Chiles picked early will be milder than those left to ripen longer in the sun.

Perennial delights

Chiles are usually grown by gardeners as annual plants and discarded after they've finished cropping in early fall. However, they are technically perennial and if you bring your plant indoors before the fall frosts arrive, you can expect more delicious chile peppers the following year.

how to grow

01 Use a knife to halve the chile or pepper, wearing gloves to protect your skin. Scoop out some of the small white seeds and place them on paper towels for an hour to dry.

02 Fill a seed tray with peat-free, general-purpose potting mix. Sprinkle the seeds evenly over the surface of the soil, leaving about ½in (1cm) between each one.

03 Cover with another ½in (1cm) of potting mix. Place in a heated propagator in a light or sunny room.

04 If you don't have a propagator, cover the tray with a plastic bag and leave in a warm, sunny position indoors. Seeds germinate quite quickly—they should emerge after about 10 days.

05 If you are intending to grow them outside eventually, the seedlings can be pricked out individually and placed into 3½-in (9-cm) pots when they have reached 4–5 in (10–12 cm) high.

06 Keep these on a sunny windowsill and when the risk of frost is over they can be planted into their final position outside, spaced 16 in (40 cm) apart.

Let it thrive

Choose a sheltered spot in full sun. Chiles can also be planted into individual 11–13-in (10–12-liter) pots, or plant three in a grow bag. In a greenhouse or on a windowsill, plant directly into 11–13-in (10–12-liter) pots when you thin out the seedlings. Taller chiles or peppers may need staking when laden with fruit. Chiles are tender plants, so make sure you have picked all those grown outside before it starts to turn cold, to avoid frost damage.

Storage

Chiles are best eaten fresh, but they can be left in the sun to dry for use at a later date. You can also freeze them. Once defrosted they lose some of their shape, but retain their heat and flavor and can be added to any dish that you want to spice up. I was surprised this year to defrost some Trinidad Scorpions, Carolina Reapers, and Chocolate Habaneros that had been in the freezer for two years, and I was still able to germinate their seed.

Hack.15
Garlic
from a clove

Garlic cloves can be planted in the garden or in pots and will each produce an entire garlic bulb for you to enjoy. Despite its association with Mediterranean recipes, garlic is Asian in origin, possibly from southwest Siberia. This means that garlic is extremely hardy, so you can even plant it out in the fall and winter, as long as there isn't snow and ice on the ground.

Garlic is sold as bulbs in the supermarket. Each bulb is made up of the individual cloves that we're all familiar with. If you stick one of these cloves in water or in soil or compost, it will propagate itself, produce new roots and shoots and become a new plant. It's amazing and so easy to do. Each clove that you plant should form into a large bulb containing lots more cloves. It's an easy way to multiply the number of cloves you have. And if you store some of the cloves somewhere dark and cool after harvesting them, you can plant them out the following season. You may never need to buy garlic ever again. As they refer to it on social media, it is a garlic infinity glitch—an endless supply of garlic.

You will need

+ Garlic bulb
+ Jar of water
+ Toothpicks (optional)

how to grow

01 Select a healthy-looking bulb with cloves that feel hard or firm. Fill a small jar with water and rest the entire garlic bulb on top, making sure the base of the bulb is above the water. If there is an excessive amount of the papery, white skin then remove it, but otherwise it can stay.

02 If the bulb is too small to rest on top of the jar, carefully push two toothpicks into opposite sides of the bulb and rest the toothpicks on the jar instead so the garlic is hovering above the water. Leave the jar on a warm, sunny windowsill.

03 After about 10 days you will notice roots emerging from the base and shoots at the top of the bulb.

04 Remove the bulb from the water and break into individual cloves.

05 Plant the cloves in a large pot or directly in the ground (see box); space them 5 in (12 cm) apart. Make sure that the top of each clove is ¾ in (2 cm) below the surface of the soil.

06 If you are growing garlic in pots, water occasionally during dry spells. Don't overdo it, because the swelling bulbs can be prone to rot in damp conditions.

Let it thrive

Your garlic cloves need a warm, sunny position to thrive and fully develop into a large bulb. Garlic also needs a light soil, so if your garden is on clay you might want to consider growing it in containers. Garlic is ready to harvest from mid- to late summer when the stems start to turn yellowish-brown and flop over. Carefully pull them up. Each original clove should have formed a new bulb of garlic. Occasionally the original clove won't have formed an entire bulb, but will instead have swollen and become much larger than when it was planted. This large clove is still fine to cook with. After harvesting, garlic bulbs will benefit from being left in the sun for a few days to cure and help develop the flavor. If you ended up with lots of bulbs you can braid the stems together and hang them somewhere cool and dry.

Hack.16
Lemongrass
from a stem

If you love Asian food, and particularly Thai cuisine, then lemongrass is an essential crop. Not only does it add a zesty, citrusy flavor to dishes, but it is also a real looker. I love its upright, chunky stems and its bluish-green, fountain-like foliage. It's an "edimental," both edible and ornamental.

Often sold in packages containing a few stems that are about 6–8 in (15–20 cm) long, lemongrass from the supermarket is easy to grow, especially if you live in a warm climate. But don't be put off growing it if you live in a cooler area, such as I do in England. Lemongrass requires a tropical climate to grow outside all year, but you too can grow it all year round if you keep it in a pot and bring it inside during the colder months, when it will make a beautiful houseplant for a sunny kitchen windowsill.

You will need

+ Lemongrass stem
+ Glass
+ Dark tape
+ 3½-in (9-cm) pots
+ Peat-free, general-purpose potting mix

Know your plants

The botanical name of lemongrass is *Cymbopogon citratus* and it originates from coastal Southeast Asia, although it is now grown in tropical areas around the world. The plant is a tender perennial grass, meaning it keeps growing year after year, but cannot survive outside in low temperatures; below 50°F (10°C) plants are liable to die.

how to grow

01 Wrap dark tape, such as duct tape, or black paper around your glass. The lemongrass stems prefer a dark environment, which encourages root formation.

02 Place the base of a stem in about ½in (1cm) of water in your glass and leave in a sunny spot for two weeks. Change the water a few times to replenish the oxygen supply, which helps promote root growth.

03 You will notice the tip of the stem starts to extend upward and you may also observe roots forming around the base, although this doesn't always happen immediately.

04 At this stage, pot your lemongrass stem into peat-free, general-purpose potting mix, making sure the part that was in water is now in the soil.

FROM THE HERB AISLE

05 Keep your lemongrass nice and warm in a sunny location and enjoy watching it grow. Water about once a week, depending on temperatures: more frequently in periods of hot weather.

Let it thrive

Keep the surface of the soil free of weeds to prevent them from competing with the lemongrass. They won't require much watering in winter, but give the plants a drink once a week during the summer growing season. They might also benefit from a general-purpose liquid food every couple of weeks. Read the instructions on the label before applying.

Hack. 17
Ginger
from roots

Ginger is one of the most flavorful ingredients you can grow at home. Not only do its chunky roots provide you with a punchy, spicy, and exotic flavor, but they make attractive houseplants too. They are perennial plants, so they'll keep cropping for as long as you take care of them. If you give the plant occasional tender loving care, you may never need to buy ginger again.

The ginger plant is grown from the underground rhizome, which is basically a fancy name for its swollen, underground roots. It is this spicy-flavored rhizome that is usually used in cooking, although their long leaves and stems can be chopped up and added to dishes too. Look for the plumpest and freshest ginger you can. If you look carefully, you will see that some of the rhizomes have small bulges or knobby bits just below the surface. These are the sections of rhizomes that you want because they are the dormant buds, which will sprout into a plant for you.

You will need

- Fresh ginger
- Sharp knife
- Small pot or seed tray
- Peat-free, general-purpose potting mix
- Perlite (optional)
- Larger pots, for repotting (optional)

how to grow

01 Using a sharp knife, carefully cut your ginger rhizomes into sections 1¼–2 in (3–5 cm) long.

02 Fill a pot or seed tray two-thirds full with peat-free, general-purpose potting mix. Perlite can be added to increase drainage, but it isn't essential.

03 Lay individual sections of rhizome on the surface of the soil in each pot. If any rhizome section has small, knobbly lumps just below the surface, make sure they are pointing upward.

04 Cover the rhizomes with potting mix to just below the top of the pot. Leave it somewhere warm to encourage the rhizomes to send out roots. Regularly check the soil and if it feels dry to the touch then give it a drink.

FROM THE HERB AISLE

05 It usually takes a year for the ginger to produce rhizomes for harvesting. Remove the plant from its pot and pull off a section of root to use in the kitchen. Then replant the ginger, adding more potting mix if necessary.

Let it thrive

Ginger plants come from Southeast Asia so they are suited to tropical climates. They are tender and won't cope with frosts or cold. If you live in a cool climate, you will need to grow your ginger indoors in a warm, sunny spot. You can put the pot outside in a sunny, warm spot during summer. Just don't forget to bring it back indoors before the temperatures take a dip in the fall.

Try turmeric

If you have success with ginger, then give turmeric a try. Like ginger, it is grown from the rhizome (root) using exactly the same technique. Not all supermarkets sell fresh turmeric, but many health food stores supply it. Look for the plumpest, freshest rhizomes. Turmeric is also a perennial, although the foliage dies back during the winter. Keep the plant indoors between early fall and late spring.

Hack.18
Cilantro
by dividing a pot

If you love Asian food, then cilantro is definitely an herb worth trying to grow. It is packed full of spicy and aromatic flavors. Whether you grow it for the bitter yet floral coriander seeds or the zesty, parsley- or celery-like stems and foliage, it is an ingredient worth including in every garden.

Buying a pot of cilantro is the horticultural bargain of a lifetime. You are getting hundreds of plants for the price of one, because there are usually hundreds of potential seedlings in there. The garden nurseries who supply pots of cilantro to supermarkets find it easiest to fill a small pot with soil and sprinkle seeds over the surface. Once all those seeds have germinated it provides the shopper with what looks like a bushy cilantro plant, but is in reality many tiny seedlings clumped together. The downside is that it isn't possible for hundreds of seedlings to survive long in such a tiny pot.

To overcome this problem, you can separate them and grow them on by "division," which means splitting or separating the root ball of the plants into smaller sections.

You will need

+ A living pot of cilantro
+ Four to six pots
+ Peat-free, general-purpose potting mix

how to grow

01 Remove the whole cilantro plant from its pot.

02 Use your fingers and thumb to carefully prize sections apart, keeping the roots as intact as possible. You should end up with four to six sections of cilantro, each with roots.

Herbs to divide

Try dividing these herbs too, which you can often find in a pot at the supermarket: mint, parsley, tarragon, and thyme.

What's in a name?

Cilantro is often called Chinese parsley, because the flavor of the leaves and stems is reminiscent of that herb. In the UK, we generally call both the edible leaves and shoots as well as the seeds "coriander." In the US the foliage is called cilantro, and the seeds are referred to as coriander.

03 Place each section of cilantro in its own pot. Top up with potting mix around the edges and firm in.

04 Leave on a sunny windowsill and keep watering (see box). You can also plant cilantro outside in spring after the risk of frost is over. Find a warm, sunny location in well-drained soil.

Let it thrive

Cilantro plants are annuals, meaning they will only live for one year before producing seeds and then dying. In hot, dry conditions they can "bolt," meaning that they grow too quickly, causing the plant to go to seed rapidly and surviving for even less than one year. If you keep the plants well watered they should survive for the majority of the summer. Keep trimming the shoots and leaves, because this will encourage more growth. If you want to collect seeds, stop trimming the plants and allow them to flower instead; the flowers will eventually turn to seed. You can either use the seeds in cooking or save them and sow again in spring for another crop the following summer.

Once you have harvested the seed, the plants can be pulled out and added to the compost heap or, if they are growing outside, left to die back in the soil. Chances are that the seeds that fall from the dead plant will germinate in the soil the following spring.

from the Fruit Aisle

Hack. 19
Strawberries
from fresh seed

Strawberries are the taste of summer—and the plants are tiny. If you have any outdoor space, from hanging baskets to tubs on the patio, you can grow them. Their trailing foliage and white flowers are beautiful in spring.

Strawberries are usually divided into two types. Larger-fruiting strawberries fruit during early summer, typically in June. Then there are the "everbearing" types, which are usually smaller and crop periodically throughout summer and into the fall. Most of the strawberries in supermarkets are June-fruiting types. Technically, a strawberry isn't a fruit, because the seed isn't contained within its flesh. Instead, the tiny, grainy black or green specks on the skin are the seeds. Not all seedlings come true from seed, so the strawberry fruit that you produce may differ from the original. Some may not ever produce fruit, but the majority will—and it's worth trying, because it's fun and will, with luck, supply you with lots of strawberries each summer for years to come.

You will need

- Strawberries
- Knife
- Paper towels and plate
- Seed tray
- Peat-free, general-purpose potting mix
- Propagator (optional, see p.13)
- 3½-in (9-cm) pots

Why is a strawberry called a strawberry?

Possibly because straw is placed under the berries as they ripen to prevent rotting on the ground. Another theory is that they were called straeberries or stray berries, referring to their "runners" or adventitious shoots, which "stray" from the plant to try to reproduce.

how to grow

01 Slice thin slivers off the strawberries. Lay them out on paper towels on a plate.

02 Leave to dry for 48 hours, then use your finger to gently rub the fruit slivers. The tiny dark seeds should detach from the dried skin. Pour the seeds onto damp paper towels.

Let it thrive

Strawberries need to be grown in a sunny position. When they start to produce fruit, it's important to place straw or other natural soft material under the berries to stop them from rotting on the ground. Birds love the fruit almost as much as humans, so you may need to think of some protection to keep them away when the fruit is ripe. One technique is to paint stones red and place them near the plants before the fruit ripens. Birds are attracted to the stones, but once they realize that they aren't edible they turn their attention elsewhere and don't take an interest when the strawberries finally ripen and turn red.

03 Half-fill a seed tray with peat-free, general-purpose potting mix. Put the paper towels with strawberry seeds on top and cover with another ½ in (1 cm) of potting mix.

04 Keep the seeds indoors on a sunny windowsill and water every few days. Be patient—the seeds can take a long time to germinate. One batch of mine took more than two months. (You can also use a propagator, see page 13.)

05 Once seedlings appear they can be pricked out and potted into individual 3½-in (9-cm) pots to allow the plants to grow and develop.

06 In spring, the strawberry plants can be planted directly into the ground, 12 in (30 cm) apart, or into hanging baskets and containers.

Hack.20
Lemon tree
from a pip

Create a touch of the Italian coast on your patio or in your conservatory with this attractive citrus plant that can supply you with lemons for your cocktails or lemonade. As a bonus, its flowers are beautifully perfumed.

You will need to be patient with this project. Lemon trees can take a number of years before they bear any fruit when grown from a pip. And there is a small chance that they might never produce anything. For this reason, I suggest sowing a few pips to increase your chances. It is fun to grow lemon trees, and even if you don't get any lemons, they are beautiful evergreen plants to have in your house and garden. I've found that while the majority of lemons from the supermarket contain pips, some don't have any. Generally the larger the lemons are, the more likely they are to contain some.

You will need

+ Lemon with pips
+ Knife
+ 3½-in (9-cm) pots
+ Peat-free, general-purpose potting mix
+ Recycled clear plastic bags
+ Propagator (optional, see p.13)
+ Pots of increasing sizes
+ Citrus potting mix (optional)

Limites, oranges, clementines, tangerines, satsumas, and most other citrus fruits can be grown from pips in exactly the same way as lemons, although finding limes with pips is difficult.

These small green fruits will ripen to yellow, providing they get plenty of sun.

how to grow

01 Cut open the lemon and remove the small white pips. Soak the pips in water for 24 hours to soften the outer shell.

02 Peel back the outer shell to reveal the smaller seed inside. You can also plant the pip with the white outer shell; it will just take slightly longer to germinate.

03 Fill a 3½-in (9-cm) pot with peat-free, general-purpose potting mix and push the seed ¾ in (2 cm) below the surface. Place a clear plastic bag over the pot to increase humidity and retain heat.

04 Place on a sunny windowsill. In winter, you may need to place it in a propagator or on a shelf near a heating vent. In a few weeks your seedling will emerge. Remove the plastic bag.

05 When it is about 5 in (12 cm) high you can plant it into a slightly larger pot. It is also worth changing from general-purpose to a specialized citrus potting mix, which will suit the lemon's needs.

Let it thrive

Lemon trees are slightly tender (vulnerable to cold), and mature plants are best kept in pots so they can be brought inside during winter months, unless your garden is very sheltered and frost-free. Nor do they like fluctuations in temperature, so when you move them indoors before it gets cold, keep them away from heating vents or they will start to drop their leaves. Feed them once a month between spring and fall with a specialized citrus food.

Know your plants

Lemons originate from the Mediterranean and are in fact evergreen shrubs: woody plants smaller than trees, which don't drop leaves over winter. Lemon plants produce small, white, fragrant flowers, which, once pollinated, will turn into the juicy, yellow fruits. Their botanical name is *Citrus* x *limon*—that "x" in the middle means that the lemons we grow today are the result of two different wild citrus plants cross-pollinating, a very long time ago.

Hack.21
Plums
from a stone

The season for plums is from mid- to late summer, when you can usually find these juicy fruits in stores. Plums grow on medium-size trees, so if you want to try this project, then you may need some space in your garden. However, you can restrict their size by growing them in containers on the patio.

Most of us think of plums as purple-reddish, egg-shaped fruits, but there are hundreds of varieties. They vary in size enormously, and come in colors including white, green, blue, yellow, gold, and black. Bear in mind that you are in for the long haul here—it can take several years for a plum to produce fruit from a seedling. Furthermore, because seedlings often vary from their parent, there is no guarantee of fruit. But it is lots of fun growing plum seedlings—they are one of my favorite fruits and, you never know, you may create an exciting new variety by sowing plum stones. Plums will benefit from other plum trees growing nearby to increase pollination, although a few can be self-fertile. Grow a few seedlings if you have room.

You will need

- Fresh plums
- Knife
- Hammer
- 3½-in (9-cm) pots
- Peat-free, general-purpose potting mix

how to grow

01 Use a knife to cut each plum in half and remove the stone. The seed is inside the stone, so carefully crack the outer shell of the stone with a hammer and remove the seed.

02 You can keep the stone intact, although germination will be much slower because the seed has to break out of the stone.

03 Fill a 3½-in (9-cm) pot with peat-free potting mix to just below the top. Use your finger to push the seed into the soil so it is ¾ in (2 cm) below the surface. Brush back soil into the hole so the seed is completely covered.

04 Put the pot outside and water regularly. Plum seeds need a cold spell to germinate (this is called stratification), so keep them outside in the cold during winter.

05 In spring your seedlings should emerge. Keep them watered throughout spring and summer, and by early fall the seedlings should be ready for planting outside in the ground, or moving into larger containers.

Let it thrive

Plums need a warm, sheltered, and sunny spot to grow in. They can be grown as a free-standing tree, but because they flower early in the season, their blossom can be susceptible to spring frosts. For this reason, it is worth planting them against a south- or south-west-facing wall to provide extra warmth and frost protection. Unlike many other fruit trees, plums don't like being pruned. It's best to do this only when the canopy is over-congested, or to remove crossing, damaged, or diseased branches. Pruning should only be done when the tree is actively growing, between spring and late summer. Pruning at other times can make it susceptible to a few diseases.

Keep it in the family

Plums belong to the *Prunus* or cherry family, and you can use this technique for other members of the family. You could also try growing cherries, apricots, peaches, nectarines, damsons, bullaces, greengages, and almonds. Peaches and nectarines will require a warm, frost-free site.

Hack.22
Apples
from a pip

There are more than 2,500 varieties of apples, yet sadly we only find a handful in the supermarket. However, by taking pips from some of these well-known names you could create your own delicious variety. Some of the tastiest apples in the world started life as seedlings. If the name includes "seedling," such as Bramley Seedling, or "pippin," such as Cox's Orange Pippin, then it almost certainly means they started off their lives as random seedlings.

Apples thrive in cool climates. They are hardy trees and will tolerate frosts and lots of rain, making them ideal candidates for growing in the garden. Commercially bought apple trees are usually grown on "dwarfing" rootstocks to reduce their size. Your seedlings will therefore grow much bigger than the trees you see in many places. However, apple trees are suitable for growing in containers, which will restrain their growth. Apple trees can take a few years to produce a crop, and like many fruit seedlings, the ones you grow won't necessarily be the same as the tree you took the pips from. Some may not even fruit at all, so sow a few pips to avoid disappointment. If they haven't fruited after five years, discard them and make room for other seedlings. You can also use this method to grow pears and Asian pears.

You will need

+ Apples
+ Knife
+ Paper towels
+ Freezer bag
+ 3½-in (9-cm) pots, plus larger sizes
+ Peat-free, general-purpose potting mix

how to grow

01 Cut an apple from stalk to tail and remove the pips (seeds) from the central core. Discard any pale or white pips, because these haven't ripened and probably won't germinate. You want the dark brown, almost black pips.

02 Wrap the seeds in damp paper towels and place in a freezer bag. Put them in a dark cupboard for a few weeks; no need for watering because the bag will retain moisture.

03 After a few weeks, remove the seeds from the paper towels. The seeds will have germinated and grown little tails. Fill a 3½-in (9-cm) pot with potting mix.

04 Use a dibble or stick to make three or four holes in the potting mix, about ¾ in (2 cm) deep, and equally spaced. Carefully place a seed in each one, and cover with more potting mix.

05 In spring, your apple seedlings should germinate. When the seedlings have reached about 4 in (10 cm) high they can be potted individually into larger pots.

06 In the fall, the apple trees can either be planted in the ground outside (although they will get big) or planted into larger outdoor containers, which will restrict their size.

Let it thrive

Apple trees do best in a sunny location in well-drained soil. They are usually grown as open-center or goblet-shaped trees, with a short trunk and then four or five fruiting branches as their main framework spanning outward. Apple trees benefit from being pruned once a year in winter. Aim to remove branches growing into the center of the tree and remove congested canopies so there is about a hand's width between each branch. This will ensure that sunlight reaches the remaining branches, which will help the fruit ripen.

Hack.23
Raspberries
from fresh seed

I remember walking into the mess room at the RHS Wisley gardens when I first started working there as a gardener in the fruit department. It was a hot summer day and there were masses of cartons of just-picked raspberries on the table, ready to go to the store. The aroma was amazing, and the scent of this fruit is still my all-time favorite.

Of course it isn't just the aroma that we love about raspberries. It is that sweet yet sharp flavor combination that is so magical. I love the way it can be added to desserts such as chocolate jelly roll or a fruit pavlova, where the acidity of the berries cuts through the sweetness of the other flavors. Yet the best way to enjoy raspberries (in my opinion) is to eat them fresh with just a dollop of crème fraîche on top. Raspberries are closely related to blackberries (brambles), and most of us know how readily they spread. This makes propagating raspberries quite easy. I use fresh raspberries from the supermarket, but I have heard that you can also use frozen ones. I must try it one day.

You will need

+ Raspberries
+ Strainer
+ Spoon
+ Plate and paper towels
+ Seed trays
+ Peat-free, general-purpose potting mix
+ 3½-in (9-cm) pots

how to grow

01 Place a handful of raspberries into a fine mesh strainer and use the back of a spoon to crush the fruit and remove the tiny seeds by pushing the pulp to one side.

02 Rinsing under water can also help separate pulp and seeds. Once you've done your best, scrape the seeds from the strainer with the spoon and leave them on a plate to dry.

03 Spread the seeds on paper towels, like jam on toast, and leave to dry for a few days. Fill a seed tray with peat-free potting mix and place the paper towels on top.

04 Cover the seeds with another ½in (1cm) of peat-free potting mix. Leave on a sunny, warm windowsill. After a few weeks your raspberry seedlings should emerge.

> ### Other berries to try
>
> Use this method for any "hybrid" raspberry such as blackberries, huckleberries, tayberries, Japanese wineberries, and loganberries.

05 Pot them into individual 3½-in (9-cm) pots filled with peat-free potting mix. Raspberries are hardy, so they can be planted out any time of year. Once they have reached 6 in (15 cm) they are ready for planting in the garden.

Let it thrive

Raspberry plants produce fruits on "canes," which look a bit like brambles. They will need a support system to keep them propped up. I plant my raspberries 8 in (20 cm) apart in rows. There are two different types of raspberries; summer and fall fruiting. The easiest way to tell which one you have is to look at where the plant is producing fruit. If it produces raspberries on canes formed in the previous year it is a fall type. If it produces berries on the canes formed on the current year's new growth, then they are the summer types. The time when they crop can also be an indicator, although this is not always accurate. Prune fall-fruiting canes down to the ground each March to encourage new canes to grow. For summer-fruiting raspberries, cut away the older fruiting canes after they have cropped (mid- to late summer) and keep the new canes produced that year, to produce fruit next year. I hope you're following this!

Hack.24
Pineapple
from a cutting

In the eighteenth century, rare and expensive pineapples were status symbols, and stone or marble replicas adorned many prestigious gateways. It showed the world that you were wealthy enough to grow pineapples in your grounds or import them from abroad. These days you don't need a team of gardeners and a pineapple pit—just a windowsill and a supermarket.

Pineapples are tender plants and you will need to grow them inside. They make gorgeous, leafy-looking houseplants, and if you are fortunate they may even reward you with fruit. Pineapples are grown by taking a cutting of the fruit's leafy growth from the top of the fruit, and I'm not aware of any other supermarket crops that can be propagated in this way. I have had varying success with pineapples: some thrived and some rotted in the water. Next time you treat yourself to a pineapple, give this a try.

You will need

- Pineapple
- Sharp knife
- Pots of varying sizes
- Peat-free, general-purpose potting mix

how to grow

01 Choose a pineapple with a green, healthy-looking top. Avoid any with brown tips to the foliage or withered leaves. Carefully twist the leafy tops and pull.

02 The whole top should come away with a small amount of flesh. Otherwise, cut the top with a knife, including about ½ in (1 cm) of the fruit.

03 Leave to dry for 24–48 hours. Peel back some of the lower leaves to reveal more of the base. I usually remove about three layers.

04 Place some potting mix in a small pot, then the pineapple. Add potting mix to just below the top of the pot. Use your fingers to firm the soil in around the pineapple.

05 Once roots have appeared you can pot your plant into peat-free potting mix in a 7-in (3-liter) pot. Keep on a sunny, warm windowsill and water every few days (see box).

Let it thrive

Keep your pineapple plant inside in a warm sunny position. It requires temperatures of 64–86°F (18–30°C), so it can be moved outside in its pot during summer. Just don't forget to move it back in when the temperatures dip. The best way to water a pineapple is by pouring water down the center of the leaves at the top of the stem. Gravity will ensure that the water reaches the necessary parts. Pineapples belong to the bromeliad family, and the rosette of top leaves is how they funnel water down to other parts when it rains. If you are lucky enough to see the pineapple start producing a fruit, then you can add a general-purpose liquid food for houseplants in the watering can (follow instructions on the label) and feed every couple of weeks.

Hack.25
Blueberries
from seed

Their delicious flavor aside, one of my favorite reasons for growing blueberries is their spectacular fall color. Forget your expensive Japanese maples at this time of year (although I do love them): blueberry, for me, is one of the best shrubs for fall color, with foliage that can offer a riot of purples, yellows, oranges, and reds.

Blueberries grow as deciduous shrubs that reach about 5 ft (1.5 m) in height. They do require an acidic soil, like the conditions that you would grow rhododendrons and camellias in. If you don't have soil with low pH in your yard, you can grow them in containers or raised beds. You will need to use ericaceous potting mix, which is formulated specifically to suit plants that prefer acidic soil. It can take a few years for blueberries to produce fruit. They benefit from other nearby blueberries to increase pollination, and therefore increase fruit yields, so it's worth planting a few blueberry seedlings.

You will need

- Blueberries
- Paper towels
- Spoon
- Strainer (optional)
- Seed tray
- Peat-free ericaceous potting mix
- 3½-in (9-cm) pots Rainwater (not tap water)

how to grow

01 Place a handful of blueberries on paper towels and use the back of a spoon to squash them. You want to separate the tiny seeds from the flesh.

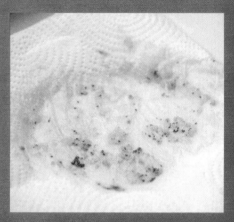

02 You can also put the squashed fruit in a fine mesh strainer and run it under the faucet to wash away some of the flesh. Leave the seeds on paper towels to dry out for 24 hours.

03 Fill a seed tray with peat-free ericaceous potting mix to just below the top.

04 Put your paper towels and seeds on top, and cover with not quite ½in (1cm) of potting mix. Leave outside; blueberries need a cold winter to enable the seeds to germinate.

05 Seeds should germinate in spring. Once the seedlings reach about 4 in (10 cm) they can be potted individually into larger containers. Blueberries can be planted into their final growing position in the fall.

Try cranberries

Cranberries can also be grown using this technique. They have the same acidic soil requirements as blueberries but need even more moisture, because their natural habitat is boggy conditions. Cranberries are much more low-growing than blueberries and have an attractive trailing habit. I've got one cranberry plant at home, in a hanging basket. I keep it slightly in the shade, so it doesn't dry out too quickly in summer, but have to keep it regularly topped up with rainwater.

Let it thrive

Blueberries prefer slightly moist conditions, so keep the plants well watered during dry periods. Use rainwater (from a rain barrel, or leave watering cans or buckets outside during rainy periods). Tap water can have a high pH (alkaline), particularly in limestone and chalk areas. Rainwater has a lower pH and is more suitable for blueberries' acidic requirements. Prune blueberries in early spring just as the buds start to appear. This helps you identify if any branches have died. Remove about one fifth of the growth. Remove older stems near ground level with a pair of loppers, leaving younger and new shoots to develop and produce the fruit. Note, although blueberries need cold to break the dormancy of the seeds (and start to germinate), if the supermarket refrigerated them, this might not be necessary.

Hack.26
Passion fruits
from seed

Passion fruits have exquisitely intricate flowers. Many of us are familiar with the purple-and-white-flowered varieties in our gardens. The fall fruits produced from these hardy varieties are barely edible, but are closely related to the tropical species that produce the delicious fruits you buy from the supermarket.

Passion flowers (*Passiflora*) are a group of perennial climbers which, as well as stunningly intricate flowers, also have attractive, deeply lobed foliage. They originate from some of the warmer regions of the Americas and many are tender, including the edible fruiting types found at the supermarket. For this reason you will have to grow your seedlings indoors, or at least under glass if you have a conservatory or greenhouse. You will therefore need enough space indoors for a large container and something for their tendrils to grip and scramble up.

You will need

- One passion fruit
- Sharp knife
- Strainer
- Spoon
- Paper towels
- Seed tray
- Peat-free, general-purpose potting mix
- 3½-in (9-cm) pots
- Large container (at least 9 inches/5 liters)

how to grow

01 Cut the passion fruit in half with a knife. Scoop out the hard black seeds. Rinse them in a strainer to remove some of the flesh.

02 Leave the seeds on paper towels to dry for a few hours. Fill a seed tray (I use a lidded fruit container) with peat-free potting mix to just below the surface.

03 Push three or four seeds into the potting mix to a depth of ¾ in (2 cm), using your finger. Fill the holes with more potting mix, and then water.

04 Close the lid, leave on a sunny windowsill, and water every few days. After a few weeks your seedlings will emerge.

05 Pot the seedlings on into individual 3½-in (9-cm) pots. When they reach about 6 in (15 cm) high, they can be potted into their final growing position; this should be a large (9-in/5-liter at least) container full of peat-free potting mix.

Let it thrive

Passion fruits are evergreen climbing plants so they will need a trellis, or support system made of strings or wires, to scramble up. Flowers and fruits should appear after two or three years if you keep the growing conditions warm and light; you could be rewarded with a small crop of passion fruits in the fall. Passion flower plants may need tying in as they grow, although they do have tendrils to help them climb. Occasionally you may need to trim or prune shoots to tidy them up and remove any surplus growth. Keep the plants well watered during hot sunny spells and occasionally give the plants a liquid food, such as tomato food, every few weeks in summer. Tomato food is high in potassium and should help your passion fruit produce flowers.

Hack.27
Mango
from a stone

Mangoes are tropical fruits that grow on large evergreen trees. Unless you are lucky enough to live in a warm area with no risk of frosts, you will need to grow a mango plant in a pot indoors.

Look for the ripest mangoes you can find in the supermarket. This usually means those with the reddest skin, although some varieties have a naturally greener color. If you gently press your finger into the surface and it feels soft, that is an indication it is ripe. The ripest mangoes will have the most developed seed, which is most likely to germinate and develop into a mango plant. I grow my mango tree in my conservatory. It is yet to produce fruit, because it is only two years old, but I'm hoping in a couple more years it may produce a crop.

You will need

- One ripe mango
- Sharp knife
- Paper towels
- Butter knife
- Freezer bag or plastic box
- Peat-free, general-purpose potting mix
- 3½-in (9-cm) pots
- Large container (at least 9 inches/5 liters)

how to grow

01 Cut the mango flesh carefully from around the large, flat stone. Give the stone a rinse and rub to remove as much pulp as you can. Then leave it to dry on paper towels for a few days.

02 Remove the seed from the stone. Carefully insert a sturdy butter knife (don't use a sharp knife in case you stab your hand)—there is usually an access point on the inside edge in the middle of the stone.

03 Twist and lever the two halves of the stone. Eventually they will come apart and you are left with the large kidney-shaped seed. Take care not to damage the seed.

04 Place the seed into a freezer bag, spray a bit of water in, and leave it somewhere warm and dark. After a few weeks, it will have germinated and sent out a sprout.

05 Remove the mango seed from the bag and push it into a pot of peat-free, general-purpose potting mix so it is about 1¼ in (3 cm) below the surface. Fill in the hole you have made with more potting mix.

06 Leave on a bright, sunny windowsill. Once the plant has reached about 4 in (10 cm) high it should be potted into a larger container. Keep potting it into larger containers as it grows.

Let it thrive

Mangoes are tender, so should be kept indoors between early fall and late spring. However, if it is manageable in size, the mango plant can be moved outside to the patio for some of the warmer months. Otherwise, keep it in a well-lit, warm room, but avoid too much direct sunlight, which can scorch the leaves through the windows. To prevent the mango from growing too tall and becoming a tree, pinch out the growing bud after a year or two. This will encourage the plant to branch outward and create a bushier, more manageable habit for growing inside. Keep the plant regularly watered and occasionally give the tree a liquid food.

Hack.28
Kiwis
from seed

Growing a kiwi from the supermarket will always hold a special place in my heart. It was my first viral post, which received more than 32 million views. To be fair, it was mainly people from New Zealand commenting about the fact I should have said kiwi fruit, because a kiwi is a bird. I do apologize to them—I wasn't suggesting we cut birds in half! In other countries, kiwi does mean the fruit.

Kiwi (*Actinidia chinensis* or *A. deliciosa* plants) is a deciduous, climbing plant originating from China. The plants are surprisingly hardy and can be grown outside in most cool climates. Kiwis can be either male or female, so to make sure you get a plant that will fruit (female), it is best to grow a few kiwis together to ensure their flowers are pollinated and they produce a crop.

You will need

- One kiwi
- Sharp knife
- Spoon
- Paper towels
- Seed tray
- Peat-free, general-purpose potting mix
- 3½-in (9-cm) pots

how to grow

01 Cut a kiwi in half. Inside you will see lots of tiny black seeds. Scoop the seeds out with a spoon.

02 Leave the seeds on paper towels to dry for 48 hours.

03 Fill a seed tray with peat-free, general-purpose potting mix and sprinkle the seeds over the surface as evenly as you can. Lightly brush the seeds into the potting mix with your hand. Leave the seed tray in a warm location indoors. Soon you will see lots of tiny seedlings emerging from the soil.

04 When they have produced their true leaves (the second set, after their first pair of leaves) you can carefully prick the seedlings out and pot them into their own 3½-in (9-cm) pots. Grow them until the seedlings reach about 8 in (20 cm) high. They can then be planted directly into the ground or into larger pots outside. Don't forget that you will need a few plants for pollination, and each plant in a pot will need a climbing structure.

Let it thrive

Plant kiwi plants about 20 in (50 cm) apart and allow them to grow and intertwine with each other. This should help encourage good pollination between male and female plants. I find it best to grow them against a wall or fence where they can scramble vertically. Alternatively, you could install two upright posts and stretch parallel rows of horizontal wires between them. Kiwis produce fruit very late in the season, so try to harvest them before the frosts get to them and damage them. You can bring them indoors and ripen them on your kitchen windowsill if they aren't quite ready and are being slow to mature. You can prune the kiwis occasionally to remove congested canopies. Cut the new shoots back to four or five buds.

Hack.29
Melon
from seed

You have two options for supermarket melons. You can either grow them from seed as healthy microgreens to add to salads, or you can grow them as plants that will produce sweet, musky fruits. Select the ripest melon you can find (any variety will do), which should have the most developed seeds. Melons should feel soft but not squishy to the touch, particularly at the top of the fruit.

Melon plants are tender and grown as annuals in cool climates. Northern areas will be better off planting them in a conservatory or greenhouse. If you are choosing somewhere outside, look for an exceptionally warm, sheltered, south-facing wall or fence to train them onto. If you want to grow melons from seed, the best time to sow seed is April or May so they are ready to go outside after the risk of frost is over. If growing for microgreens, sow seed at any time, although they make their best growth through the spring and summer. I have sometimes found melon microgreens can taste slightly bitter if you don't harvest the leaves when very young.

You will need

- One ripe melon
- Knife
- Spoon
- Strainer
- Paper towels
- Seed tray
- Peat-free, general-purpose potting mix
- 3½-in (9-cm) pots

Grow this tender fruit under cover unless you have a very warm and protected garden.

Melons will climb using tendrils to cling on to any support they find.

how to grow

01 Slice the melon in half, scoop out some of the seeds with a spoon and rinse them to remove any residual fruit flesh.

02 Spread the seeds out on paper towels for a few hours to dry.

03 Fill a seed tray with peat-free, general-purpose potting mix. Sprinkle the seeds over the top, cover with more potting mix and leave on a sunny windowsill. Water regularly and soon you will see tiny seedlings emerging. These are your microgreens.

04 If you just want microgreens, trim off the young growth with scissors once they reach ¾–1¼ in (2–3 cm) high and use immediately. They will occasionally regrow, giving you a second or even third flush a few weeks later.

05 If you want to grow melons, leave the seedlings to grow to about 2½ in (6 cm) and then carefully prick out the best ones. Pot them individually into 3½-in (9-cm) pots.

06 Keep them on a sunny windowsill. Once the risk of frost is over, they can be planted outside.

Let it thrive

Melons shouldn't be planted outside until the risk of frosts are over. They can then be potted into large containers or grow bags. You can fit two plants per grow bag. Melon plants need a support structure to climb up and the central stem should be trained up the center of it, with laterals (side shoots) spread out to either side. When melons ripen they may need something to support their weight, such as a "hammock" made from old mesh bags or even old socks (probably best to wash the socks first!).

Hack. 30
Lychees
from stones

These unusual-looking fruits are about the size of a gooseberry with a red, textured outer skin. Inside is edible, soft, succulent flesh and a large reddish-brown stone. They are fairly easy to get to germinate and with a bit of patience and care will eventually reward you with an attractive evergreen tree, suitable for growing indoors or in a greenhouse.

Lychees (*Litchi chinensis*) originate from the warmer regions of southern China and prefer tropical or subtropical conditions. They can grow into large evergreen trees in their natural habitat but their size can be reduced by growing them in pots. Patience is the name of the game when growing a lychee from seed. They are slow growing. I've a two-year-old tree that is still only about 6 in (15 cm) high, which I keep on my kitchen windowsill. It will probably be a few years before I can expect to get any fruit from it. Good things come to those who wait, apparently, so I'm looking forward to one day being able to harvest my own delicious lychees. Lychees are available from supermarkets for most of the year, usually in a package of about 10 fruits.

You will need

- Lychees
- Sharp knife
- Jar of water
- 3½-in (9-cm) pots
- Peat-free, general-purpose potting mix
- Propagator (optional, see p.13)

The lychee is slow growing, but you can enjoy it as an indoor plant while waiting for fruit to grow.

how to grow

01 Peel the red or green outer skin away from the lychees. Despite its rugged appearance, the skin peels easily by hand.

02 The almost rubbery-textured flesh is the fruit. Remove this, using a knife or your nails, to reveal the reddish-brown stone (seed) beneath.

03 Soak the stone in a jar of water for a few hours to soften the surface. This makes germination easier.

04 Fill 3½-in (9-cm) pots with peat-free, general-purpose potting mix and sow a stone in each one. Use your finger to push the stone about ½ in (1cm) below the surface and then fill in the hole.

05 Leave on a warm, sunny windowsill, or use a propagator if you have one.

06 After a few weeks you should see your lychee germinating. When it is about 4 in (10 cm) high it can be removed from its pot and planted in a larger container.

Let them thrive

Keep your lychee indoors for most of the year. During the warmest months you can move it out onto a sunny patio, but don't forget to bring it back indoors before the temperatures start to drop in late summer or early fall. Lychees require very little pruning. Just occasionally cut back any wayward branches or if you wish to improve its shape in spring. Lychee trees produce clusters of fruit. Harvest them by removing the entire cluster with pruners. An indication that they are ripe is that their skin turns red and they start to swell. Unlike many other fruits, lychees will not continue to ripen after they are picked, so it is important to regularly test for ripeness every few days and harvest once they are ready.

Hack.31
Pomegranates
from seed

Despite originating from warm places such as Iran, Iraq, and Northern India, pomegranates are relatively hardy and can be grown outside in milder areas. I grow them outside in Devon, England, and they produce juicy red fruit about the size of an apple.

Pomegranates (*Punica granatum*) are attractive deciduous shrubs which grow up to about 6½ ft (2 m) in height, although you can restrict their size by keeping them in a pot. A pomegranate will need a warm location to bear flowers and fruit. In cold areas, they can be grown in conservatories or well-lit indoor spaces. If you aren't sure whether your pomegranate is warm enough outside, try growing it in a pot and in winter moving it to a sheltered position such as a porch and draping a fleece over it. Pomegranates are packed full of plump, ruby seeds, and they seem to propagate very easily. I often end up with lots of seedlings, which I give away to friends.

You will need
- Ripe pomegranate
- Sharp knife
- Paper towels
- Jar of water
- 3½-in (9-cm) pots
- Peat-free, general-purpose potting mix

Did you know?
Despite an apple being depicted in much Renaissance art showing Eve tempting Adam in the garden of Eden, in the original story it was probably a pomegranate. The word pomegranate means "apple with seeds," from the Latin for apple, *pomum*, and seeds or grain, *granatum*, hence the confusion.

how to grow

01 Cut open the pomegranate and remove a few of the red seeds. They are covered with juicy flesh, which needs removing. You can do this by rubbing the seeds with paper towels (it can be messy) and then rinsing them. I like to pop them in my mouth and suck the flesh off the seed.

02 Put the seeds into a jar of water for 24 hours. This softens the outer surface of the seed, encouraging it to germinate more easily.

03 Fill 3½-in (9-cm) pots with peat-free, general-purpose potting mix.

04 Push three or four seeds into the potting mix, ½ in (1cm) deep.

05 Leave on a sunny windowsill and keep well watered.

06 The seeds should germinate within a month. When they are about 4 in (10 cm) high they can be carefully pricked out and potted into individual pots.

Let it thrive

Plant pomegranates directly into the ground. They require well-drained soil and full sun. Alternatively, grow in a large container indoors. Ensure the plants are kept watered during the summer and keep the surrounding soil free from weeds. Give plants in containers a general-purpose liquid food in summer when the flowers appear and continue feeding until fall. Fruits should appear in late summer. Harvest when the surface of the pomegranate feels firm, but gives slightly when pressed. Pomegranates are fairly low-maintenance shrubs and grow reasonably slowly. They don't generally require much pruning, unless there are stray branches to remove or you need to reshape them.

from the Dry Goods Aisle

Hack.32
Chamomile
from a teabag

Many herbal teas such as peppermint, sage, lemon balm, rooibos, and even traditional tea (*Camellia sinensis*) are made from leaves, but chamomile tea is made from the flowers, and during the commercial drying process many of these flowers turn to seed. Great news if you want to grow your own chamomile at home, because there are lots of seeds in just a single teabag.

Chamomile is a low-growing perennial with small, daisylike, fragrant white flowers. There are two types: German chamomile, *Matricaria chamomilla*, and Roman chamomile, *Chamaemelum nobile*, and the flowers from both can be used to make tea. Most commercial teabags will contain Roman chamomile; German chamomile is an annual, which means you'll have to sow each year for a continual supply. While chamomile is easy to grow, I must admit I've had varying success with different brands of teabag, and you may therefore want to experiment with a few. Teabags can be sown indoors at any time of year, though they might need a propagator or heat mat if sowing in winter (see pp.12–13), to stimulate germination and growth.

Try these

Other dried flowers such as rose hip, echinacea, jasmine, lavender, and hibiscus can also be used.

You will need

+ Chamomile teabags
+ Small pot or seed tray
+ Peat-free, general-purpose potting mix
+ Vermiculite (optional)
+ Dibble or pencil, for pricking out
+ Larger pots, for growing on (optional)

how to grow

01 Fill a pot or tray with peat-free, general-purpose potting mix. Tear open a chamomile teabag and sprinkle the contents over the soil. Lightly brush the tea on the soil surface so it becomes slightly embedded.

02 You can cover the potting mix with a thin layer of vermiculite if you have it, to help retain moisture and prevent competition from weeds, but it isn't essential.

03 Water, then leave somewhere warm and sunny such as a kitchen windowsill. The seeds should germinate within 14 days.

Chamomile lawn

You can also sow teabags directly outdoors in spring to make a chamomile lawn or mini wildflower meadow. Chamomile requires well-drained soil, so add a bucketful of sand per square yard and dig it into the top 2 in (5 cm) of soil. Sprinkle about 10 teabags per square yard over the surface and very lightly rake it in. You may need to protect emerging seedlings from birds.

04 Once seedlings are about 2 in (5 cm) high, prick them out. Hold the main stem between your fingers while inserting a pencil or dibble into the potting mix, and ease out the roots as you gently lift the seedling.

05 Plant seedlings individually into pots indoors or directly into borders, beds, and mini wildflower meadows outside.

06 Flower heads should appear during the summer. Pick these and allow to dry to make your own fragrant herbal tea. Let a few flowers go to seed so you can collect them and sow again next spring, if you like.

Let it thrive

Keep soil free from weeds to ensure they don't smother the chamomile and steal nutrients. The easiest way to do this is to use a hoe between the plants, taking care not to damage the roots of the chamomile. Rake up the weeds and compost them. Perennial chamomile can be cut back in early spring to encourage new growth and flowers for the following year. Annual chamomile can be pulled up in the fall and resown with seeds the following spring.

Hack.33
Lentils
from dried seed

Lentils are packed full of protein and are easier to grow than you may think. I've managed to grow large crops in my garden not only as microgreens but also for the actual pods containing the lentils. Drying the lentil crop means you can store them to cook with and have some to sow the following spring. Dried lentils can be stored for up to two years.

Lentils can be grown from the dried lentils (*Lens culinaris*) that you can buy from the supermarket. The different types include brown, green, yellow, and puy. All of these are fine to use for propagation (don't buy the split ones, which won't work). You can grow lentils as microgreens on a windowsill indoors, wherever you live. However, to grow them as fully fledged plants for their pods (which contain the lentils) requires a favorable climate and a sunny, sheltered position. I can grow them outside in South Devon, England, but in cooler areas you may struggle unless you have a polytunnel or a warm spot indoors. You could try growing them in pots outside and, if the pods haven't developed before the frosts arrive, move them indoors.

You will need

- Dried lentils (any type except split)
- Jar or bowl of water
- Strainer
- Shallow tray with drainage holes
- Peat-free, general-purpose potting mix

how to grow

01 Soak a handful of dried lentils in a dish or jar of water overnight. Drain the lentils in a strainer.

02 Keep the lentils in a jar of water in a light room for a few days. Every other day, drain and rinse the lentils and place back in the jar with fresh water.

03 After a week or so the lentils will sprout. You can either eat them at this point as a healthy snack, or plant them outside individually to produce lentils (see step 6).

04 Alternatively, fill a shallow container with peat-free, general-purpose potting mix. Make sure it has drainage holes. Sow your soaked lentils fairly densely on the surface.

05 Leave on a sunny windowsill. After a few days they will start to germinate. Use scissors to harvest the top growth when it is ¾–1½ in (2–4 cm) high if you want microgreens.

06 If you want to grow them outside for lentils, wait for the seedlings to get to about 4 in (10 cm). Plant them outside in late spring after the risk of frost is over. Carefully prize the seedlings away from the tray of soil (I use a spoon), preserving as much root as possible, and plant them 4 in (10 cm) apart, in rows 8 in (20 cm) apart.

Let it thrive

Lentils grow to about 16 in (40 cm) in height so will probably not need a support structure. If you find they are flopping onto the ground you can push small posts into the ground at the end of each row, and run two rows of parallel string along them at 8 in (20 cm) and 16 in (40 cm) in height. Tiny pea pods appear 80–100 days after sowing. Harvest these and remove the lentils from the tiny pods to cook with. Alternatively, you can remove them from the pods and leave them on trays to dry indoors somewhere warm. If you're guaranteed a warm and dry late summer you can leave them on the plant in their pods to dry. Lentils are annuals, so you will need to sow seeds each spring.

Hack.34
Peas
from dried seed

One of my favorite crops to eat directly from the garden is fresh peas in their pods. I grow them regularly, propagated from the cheap boxes of dried pea seeds you find in the supermarket. The peas they produce are just as reliable and good as anything I've grown from packets of seeds bought from a garden center.

Peas (*Pisum sativum*) are climbing annual plants with succulent pulses produced inside the pods. If you want an additional gourmet treat, harvest the young shoots, leaves, and tendrils when they first appear. Peas are reasonably hardy plants and will tolerate fairly cold conditions outside. I start sowing peas in early spring outside, and then sow regularly every two or three weeks afterward for a succession of pea harvest from mid-spring through to late summer. Peas don't like to have their roots disturbed once they have germinated, so I've found it best to grow them in toilet paper tubes. The tubes can be planted directly into the ground, without having to remove the seedlings. The cardboard just decomposes. The length of the cardboard tubes also encourages the peas to produce long roots, which they prefer.

You will need

- Dried peas
- Jar or bowl of water
- Strainer
- Toilet paper tubes
- Tray or container
- Peat-free, general-purpose potting mix

Peas do best with long shoots, which you can encourage with a toilet paper tube.

how to grow

01 Soak a handful of peas for a few hours in a jar of water, then drain in a strainer.

02 Put the toilet paper tubes on a tray and fill with peat-free, general-purpose potting mix.

03 Sow one pea per tube. Push the pea about ½ in (1 cm) into the soil with your finger and fill in with more potting mix.

04 Keep the cardboard tubes on your windowsill for a few weeks until the seedlings have reached about 5 in (12 cm) high.

05 Your pea plants are then ready to be planted outside in a sunny position in well-drained soil. Ensure the toilet paper tubes are entirely buried. If the top of the roll remains above ground it can act as a wick, sucking up the moisture from the roots.

06 Peas are ready when the pods look swollen. Break a pod open; you should see good-size, fully formed peas. If they look small and misshapen, leave any similar-size pods on the bush longer. Pick regularly to encourage the plant to produce more.

Let it thrive

As the season gets warmer, I start to sow peas directly into the ground instead of indoors. I make a shallow trench with a hoe and then sow peas in a zigzag pattern along the row, leaving about 5 in (12 cm) between each seed. These climbers need support as they grow. You can use twigs for the pea plants to scramble up; simply stick them into the ground at regular intervals along the row. When the plant has finished producing pods, it can be cut down to ground level. You can leave the roots in the ground if you don't need the space for anything else. Just chop through the ground with a spade. Because they are part of the legume family, peas "fix" nitrogen, a key nutrient for plant growth, in the soil, meaning your soil will be fertile next time you plant another crop in the same place.

Hack.35
Mustard
from the whole spice

There are two very good reasons for growing mustard seeds bought from the supermarket. The first is that they are fast growing and quickly produce spicy, flavorful leaves that are perfect for adding a bit of excitement to salads. The other reason is that you can let the pretty yellow flowers go to seed and harvest them to make your own spicy mustard.

Mustard seed is dirt cheap to buy from the spice aisle of the supermarket. If you buy one jar of seeds, you should never have to buy a jar of mustard again! I grow mustard plants in large pots on my patio. You can also sow them directly in the ground—sprinkle them over cultivated ground and they will quickly grow, flower, and seed in a short period of time. I must warn you that mustard is what I call a "keeper." Once you've got it in your garden, you might never get rid of it again. I love it seeding everywhere, but if you want finely manicured flower borders, you will need to keep an eye on it and try to grab the seed before it drops onto the ground.

You will need

- Large pot with drainage hole (optional)
- Bricks (optional)
- Bits of broken pots (optional)
- Dried mustard seeds
- Peat-free, general-purpose potting mix
- Watering can

how to grow

01 Place broken pots or stones in the bottom of the pot to prevent potting mix from washing away when you water the plants or when it rains. Fill the pot with potting mix almost to the top.

02 Lightly water the potting mix using a watering can to provide a moist surface for the seeds to settle on.

Let it thrive

If you are not sowing directly into the ground, place your large pot on bricks in a sunny, sheltered position. The bricks will help water drain out of the bottom of the pot. Keep your mustard plants well watered during dry periods. The flowers will turn to seed, forming thin green seed pods. The mustard seeds are ready when the pods turn from green to brown and take on a papery texture. Pick the seed pods regularly and store them in a box or container. When you've got enough pods to make the job worthwhile (because it is quite slow), break them open and collect the tiny black seeds inside. Store them in plastic containers until you are ready to make a batch of mustard. The seeds will last for years, so you can make just enough mustard as you need it, knowing you can make more with your stored seeds when you run out.

FROM THE DRY GOODS AISLE

03 Pour your mustard seeds into your hand and sprinkle over the soil. Brush over another ½in (1cm) of potting mix and lightly water again.

04 Mustard grows quickly, so you can expect to see shoots appearing after a few days. You can pick a few of the leaves when they appear, for salads, but leave enough for the plant to produce flowers.

Homegrown wholegrain mustard

Add equal parts of mustard seed, vinegar (I like apple cider vinegar), and water to a jar. Mix, then leave to settle and partly ferment for about 48 hours. Strain the mustard seed, saving the liquid for later. Blend the mustard seed in a food processor. It should take on a lovely yellow color. Add some of the soaking liquid to give the consistency you require, and blend again. Spoon your mustard into small spice jars. Mustard can be stored in the fridge for three to six months, and sometimes much longer.

Hack.36
Chia
from seed

Chia is an annual herb related to sage and salvias, and has numerous health benefits. I grow chia seeds to produce microgreens. It's quick and easy and requires next to no space, just a sunny windowsill. Because chia microgreens are so fast growing, you can sow them every few days to give you a continual supply.

Packages of tiny black chia seeds can be found in supermarkets and health food stores. Most people sprinkle the seeds into food and drinks as they are. However, it is far more fun (and just as nutritious apparently) to grow them as microgreens. If you have childhood memories of growing cress on paper towels, then this technique will be familiar to you. In mild areas, chia seeds can be sown outside so you can harvest the seeds at the end of the season. I'm able to do this in southern England, but it might not be so easy in cooler regions.

You will need

+ Chia seeds
+ Paper towels
+ Plate

how to grow

01 Place a sheet of damp paper towel on a plate. Sprinkle chia seeds liberally over the paper.

02 Place on a sunny windowsill. And add a light splash of water every two or three days.

Let it thrive

Chia plants grow to about 3 ft (1m) high and produce spikes of blue flowers, which will eventually turn to seed. The plants are an attractive addition to the garden, regardless of the benefits of their seeds. If you want to try growing them outdoors you will need about one square yard of well-drained soil in full sun. Cultivate the ground, removing any weeds that might compete with the seedlings when they emerge. Sprinkle a small handful of chia seeds over the area, as equally distributed as you can, and rake them in. Lightly water using a watering can. Harvest the seeds on a dry, sunny day and store them somewhere dry and cool.

03 After three days you will notice the seeds starting to germinate. The microgreens are ready for harvesting when they are about 1¼ in (3 cm) high.

04 Trim off the shoots and new growth using clean scissors and add to your favorite savory dishes. If you are lucky, the microgreens will regrow after harvesting.

Hack.37
Chickpeas
from dried seed

I was able to make my own hummus last year with the plants I grew from dried chickpeas I bought from the supermarket. It was a lovely surprise, because I wasn't sure if they would grow and ripen in my garden. But I had a fairly large crop, and I will now make sure I set aside space to grow them each year.

Chickpeas are annual plants from the legume, or pea and bean, family. You will need to grow them outside somewhere (unless you have a greenhouse or conservatory), because they won't fit on a standard windowsill. As with all pulses, the peas themselves are literally the seeds. Small pods, which carry the peas inside, will appear on the plants. You can harvest the chickpeas when they are still green, but most people like to leave the peas to turn brown so they can be used for hummus.

You will need

- Dried chickpeas
- Paper towels
- Strainer
- 3½-in (9-cm) pots
- Peat-free, general-purpose potting mix

how to grow

01 Soak a handful of dried chickpeas for 24 hours. Drain and leave to dry on paper towels.

02 After a day or two you will notice the seeds are starting to form "tails" (small shoots). This means they have germinated and are ready to be planted.

Let it thrive

Chickpeas only grow to about 20 in (50 cm) high, so they don't need support. Keep the area around your chickpeas free of weeds because they will compete for light, nutrients, and moisture, and water the plants if you experience an exceptionally dry spell during the summer. The chickpeas should be ready between mid- and late summer, depending on when you planted them. If you are expecting a dry summer, then leave the peas on the plant for as long as possible. If it is wet, you may need to harvest the pods, remove the seeds, and spread them on a tray to dry indoors. Dried chickpeas should last for a year or two. Save some for sowing again the following year and use the remainder for cooking.

FROM THE DRY GOODS AISLE

03 Sow the chickpeas in 3½-in (9-cm) pots and grow on a sunny windowsill until late spring. Shoots should emerge in 14–21 days. You can keep growing them indoors in containers.

04 Alternatively, once the seedlings are 4 in (10 cm) high, and the risk of spring frosts has passed, you can prick out and sow the chickpeas directly into the ground in rows, 6 in (15 cm) apart.

Grow your own hummus recipe

Soak 14oz (400g) dried chickpeas overnight in water. Cook in a saucepan full of water until they're soft. Take the pan off the heat. Agitate the chickpeas in the water using a spoon or spatula, which should remove most of the outer skins. You may need to peel the last few by hand. Drain the chickpeas, reserving some of the cooking water. Put them in a food processor with 4oz (120g) tahini, the juice of 2 lemons, and 2 peeled garlic cloves (no need to chop the garlic). Add a pinch of ground cumin, if you like. Blend the ingredients until the hummus is smooth. If you prefer a slightly thinner texture, add some of the cooking water. The hummus will keep, in a sealed container, for a few days in the fridge.

Hack.38
Sunflowers
from seed

Sunflowers originate from the warm climates of the southern US and Mexico, yet do amazingly well in cooler climates too. They are hardy annuals and must be one of the most vibrantly colored flowers out there. When grown en masse, the giant yellow flowers provide the most spectacular golden display in the garden. I love them, and they really do feel like a ray of sunshine. They are fast growers and a great way of getting additional height into the garden with minimum effort.

I buy raw sunflower seeds that have been removed from their husks, but seeds still contained in this outer shell are fine too. Sunflowers grow easily from the seeds, and this hack is so much cheaper than buying from plant nurseries. Although your sunflowers won't be a named variety, all the sunflowers I have grown from the supermarket have been stunningly beautiful. If you don't have space, then you can grow them as microgreens. Their tiny young shoots and emerging miniature leaves are delicious and very healthy. To grow microgreens you don't need much space, just a spot on a sunny windowsill. Use the method for chia seeds (see p.166).

You will need

- Sunflower seeds
- 3½-in (9-cm) pots
- Peat-free, general-purpose potting mix
- Paper towels
- Spray bottle

Fast-growing sunflowers are a great way to add drama to the garden.

These hardy plants do well in cool climates, despite their warm-climate roots.

how to grow

01 Fill a pot with potting mix to just below the top.

02 Use your finger to push a single sunflower seed ¾ in (2 cm) below the surface.

03 Backfill the hole with the surrounding potting mix. Water the pot and leave on a sunny windowsill.

04 Plant outside when the sunflower is about 4 in (10 cm) high.

05 If you want to grow them as microgreens, sprinkle the seeds onto paper towels and mist with water. Keep on a sunny windowsill, watering every couple of days.

06 After about a week the sunflowers will have germinated. The edible young shoots can be harvested with scissors, cutting at the base of the individual plants.

Let it thrive

Sunflowers can reach more than 6½ ft (2 m), so give each plant plenty of space—ideally about 16 in (40 cm) between each one. Individual plants may also need staking with a sturdy cane to prevent their weighty flower heads from flopping over. Once they have finished flowering, you can leave the flower heads for the birds to feed on. Don't forget to collect a few seeds yourself, so you can do the same again the following year.

Hack.39
Hazelnuts
from nuts

Hazelnuts are deciduous, small trees or bushes that make an attractive feature in any garden. Not only do they produce one of my favorite nuts, but they also produce pretty, long catkins in winter and early spring. They are a useful plant as well, because their flexible branches can be used for training and supporting plants. Hazels are fully hardy and need to be grown outside.

Hazelnuts can be grown in large pots on a balcony or patio courtyard, but you will need some sort of outside space if you don't have a yard. The seeds benefit from stratification, which means providing the seed with a cold environment for a few weeks to break its dormancy and prompt it to germinate. If you are starting in spring, you could place them in the fridge for few weeks to chill, but chances are the store you bought them from will have already had them in a cold store, so it probably isn't necessary. Hazelnuts can be grown as trees, or they can be coppiced each year to produce young, whippy stakes and canes. Coppicing involves cutting the growth of the tree back to near ground level to encourage new, healthy and vigorous shoots.

You will need

+ Raw hazelnuts
+ Nutcracker or hammer
+ 3½-in (9-cm) pots
+ Peat-free, general-purpose potting mix

Spring catkins are a bonus of growing your own hazelnuts.

You can grow a hazelnut tree, or cut it back every year to grow a crop of canes to use in the garden.

how to grow

01 Break open the outer shell of a hazelnut using nutcrackers, or gently tap it with a hammer, taking care not to damage the nut inside. Sometimes supermarket hazelnuts also have an outer husk, which should be removed.

02 Repeat to get two or three nuts. Fill a 3½-in (9-cm) pot with peat-free, general-purpose potting mix.

Let it thrive

Hazelnuts produce both male and female flowers (the catkins) on the same bush. They are therefore monoecious, meaning it is possible for them to pollinate themselves, but crops will be better if you plant more than one bush in the area. Hazelnuts are wind pollinated, so if you are lucky and have a large yard, you can plant a few in a grid to ensure maximum pollination instead of planting a single row of bushes (by the way, if you have a few hazelnut trees in your yard, it is known as a "plat" rather than an orchard). Plant the bushes 16 ft (5 m) apart. Prune hazel trees when they are dormant in winter. If you are pruning for hazelnuts, then cut back the new growth to a few buds. If you are wanting to coppice the bush then cut down to near ground level every couple of years to encourage fresh, new shoots.

03 Push the nuts below the surface with your finger, to a depth of ¾ in (2 cm). Leave outside over winter. In spring the seeds will germinate.

04 When your hazel plants are about 6 in (15 cm) high they can be planted outside in a container or directly into the ground. They require moist but well-drained soil in full sun.

Hack.40
Quinoa
microgreens from seed

Quinoa is an edible seed, high in protein and fiber and therefore considered by some to be a superfood. It's closely related to amaranth and even the weed "fat hen." It has been grown as a staple crop in South America for thousands of years, but it is a fairly new (and trendy) ingredient elsewhere in the world

Because quinoa is a seed and not a grain, when you buy a package of quinoa from the supermarket you are actually buying a package of seeds. This is why they are so easy to grow! Quinoa is an herbaceous perennial, and although most people will grow it for its seeds, it is also possible to grow it as a microgreen. For microgreens, you will be harvesting the tiny emerging shoots and leaves while they are very young and tender.

You will need

- Package of quinoa
- Shallow container with drainage holes
- Peat-free, general-purpose potting mix
- Propagator (optional, see p.13)

how to grow

01 Fill a container or a seed tray with peat-free, general-purpose potting mix to just below the top. Lightly sprinkle the seeds over the surface.

02 Cover the seeds with a tiny amount more of potting mix— literally 1/32 in (1mm) and no more. Give the seeds some water to encourage germination.

03 Leave on a warm, sunny windowsill. If you have a propagator, put your container in that.

04 Water the plants every couple of days. You should see emerging shoots after 10 days or so.

FROM THE DRY GOODS AISLE

> ### Good seed
>
> Microgreens can also be grown from chia seeds (p.166), sunflower seeds (p.174), and melon, cucumber, and broccoli seeds.

05 Harvest the microgreens when the emerging seedlings are 1¼–1½ in (3–4 cm) high. Use scissors to cut the seedlings at their base and add the microgreens to salads.

Let it thrive

Microgreens are fairly short lived, but they will sometimes regenerate after harvesting if you keep them watered and in sunlight. The next crop or two won't be as vigorous as the first. When they fail to re-shoot, put the contents of the container in the compost and start again with fresh seeds and potting mix. It is quick and easy to sow a fresh batch to keep you regularly supplied with these nutritious young shoots.

Glossary

Annual Annuals live for just one year before setting seed and dying.

Adventitious roots A preliminary set of roots that plants often send out to look for moisture and nutrients so they can sustain themselves and start growing. They also send out adventitious roots during periods of stress when deprived of the necessary nutrients and water for growing.

Anther Part of a flower's male reproductive organs, which contains the pollen.

Anthocyanin A water-soluble flavonoid and antioxidant that is found in red, blue, and purple fruits and vegetables.

Biennial A plant that lives for two years before setting seed and dying.

Blanching Covering emerging stems so they remain white (see p.39).

Bromeliad A diverse group of tender plants grown for their flamboyant and colorful displays. They usually have strap-like leaves and although most are grown as ornamentals some, such as pineapple, are grown for their fruit.

Catkin Many trees and shrubs produce pollen on long, dangling flowers known as catkins; a popular example is hazel bushes. They rely on the wind to blow the pollen onto female flowers (which can also be catkins) in order to be pollinated.

Cloche Cloches are used to cover plants and protect them from the cold, often at the start or end of the season to extend the growing period. Although traditional cloches are bell shaped ("cloche" is French for "bell") and placed over individual plants, they are now more often tunnel shaped to protect entire rows of crops.

Compost A medium used to grow plants in. It is usually a mix of organic material that has been rotted down and decomposed. See page 11 to make your own at home.

Crocks Broken bits of pottery put in the bottom of a plant pot over the drainage holes prior to adding potting mix. The theory is that it will prevent the soil washing away when plants are watered without impeding the drainage. Often stones or even coffee filters are used as an alternative for crocks in the bottom of the pots.

Dibble A stick used to create a hole in the soil to plant seeds, usually the width and length of a pencil. Gardeners often use a pencil or pen as a substitute; I often use my finger to create a hole.

Die back Some plants start to die back from the tips if they are underfed, lack water, or have a disease.

Dioecious With the male and female parts appearing on separate plants.

Ericaceous potting mix An acidic potting mix required by some plants, such as blueberries and cranberries.

Filament The lower section of the male part of the flower that supports the anther (containing the pollen).

GLOSSARY

First leaves Most plants produce an initial pair of leaves (called cotyledons) that are different to the "true" leaves that appear later when the plant is established.

General-purpose potting mix Organic material suitable for most gardening tasks such as filling containers. A good option if you only want to buy one type of potting mix.

Germination The point at which the seeds first develop into growing plants, or seedlings.

Hardy Used of plants that will tolerate cold temperatures and frosts and can be left outside over winter.

Hilling up Piling soil around emerging shoots as they grow (see p.42).

Legume A group of plants belonging to the pea and bean family, including fava beans, lentils, chickpeas, and many more.

Lobe Some leaves have a curvy shape, and the parts of the leaf that bulges out is said to be lobe-shaped.

Monoecious With both the male and female parts of the flower on one plant.

Node The fancy name for a bud, which is where emerging shoots or leaves will grow from. The space between each node is called an internode.

Perennial Plants live for varying lengths of time; and perennials will live for a few years.

Perlite Tiny, lightweight, round, white pieces of volcanic glass that have been heated to a high temperature. Commonly added to potting mixes to increase drainage and aeration.

Propagator A tray used to encourage seeds to germinate (see pp.12–13).

Pricking out Removing new seedlings from where they were sown, if they've germinated among lots of other seedlings, and planting them in individual pots or in the ground. This ensures they have enough space to grow.

Rhizome A type of underground storage vessel, similar to a root, which supplies the plant with nutrients.

Seed starting mix A friable mix, easy to sow seeds into. It will contain the appropriate nutrients required for seed germination.

Stratification A cool period of dormancy required before some seeds germinate. This is a natural defense technique for plants in the wild that prevents seeds germinating in the fall when tender young seedlings might not survive the winter. You can fake the stratification process by sticking seeds in a fridge for a few weeks before sowing.

Tender Plants that can't cope with frosts or cold temperatures are called tender. There are various levels of tenderness, and in some regions, zones that denote which plants will survive depending on their hardiness or tenderness.

True leaves The leaves that appear once a plant is establishing itself. They closely resemble the leaves on the mature plant, hence the name.

Vermiculite Tiny, pale, flake-like minerals with excellent moisture-retentive qualities. It is usually placed over soil when propagating seeds and cuttings to retain moisture and prevent competitive weeds germinating.

Index

adventitious roots 96, 186
almonds 107
apples 108–11
apricots 107
Asian pears 108
avocado 32–35

banana skins 16–17
beets 59
bell peppers 74–77
blackberries 115
blanching 37, 39, 186
blueberries 17, 120–23
bok choy 63
brassicas 60–63
broccoli 63
bromeliads 119, 186
bullaces 107

cabbage, red 60–63
carrot fly prevention 23
carrots 56–59
celery 36–39
chamomile 150–53
cherries 107
chia 166–69
chickpeas 170–73
chiles 74–77
Chinese leaf cabbage 63

cilantro 90–93
clementines 101
cloches 186
cloches, milk cartons 13
coffee filters 17
coffee grounds 17
compost 11
containers
   see also milk cartons
   drip trays 15
   homemade propagator 13
   recycled 7
coppicing 178
corn 50–51
corn on the cob 48–51
cotyledons 187
cranberries 17, 123
crocks 17, 186

damsons 107
dividing plants 92–93
drip trays 15

echinacea 150
ericaceous potting mix 120, 186

freezing
   chiles 77

green onions 23
garlic 78–81
ginger 86–89
green onions 20–23
greengages 107
greenhouses, mini
  homemade 13
growing conditions 10–11

hazelnuts 178–81
hibiscus 150
hilling up 42, 186
huckleberries 115
hummus 173

Japanese wineberries 115
jasmine 150

kale 63
kiwis 132–35

labels for plants 14
lavender 150
leaf mold 17
leaves
  first 187
  true 187
leeks 52–55
lemongrass 82–84
lemon tree 100–103
lentils 154–57
lettuce 24–27, 63
light 10–11
limes 101

loganberries 115
lychees 140–43

mangoes 128–31
melons 136–39
microgreens
  broccoli 185
  chia 166–69
  cucumber 185
  lentils 154–57
  melons 136–38
  quinoa 182–85
  squash 64
  sunflowers 174, 177
milk cartons
  cloches 13
  seed pots 14
  watering cans 14
mint 70–73
mustard 162–65

nectarines 107
nettles 16
nitrogen 16

onions, carrot fly
  prevention 23
oranges 101

pallets, raised beds 15
passion fruits 124–27
peaches 107
pears 108
peas 158–61

peat, not using 11
peppermint 70
peppers (bell and chile) 74–77
pesto 37
pineapple 116–19
plant food 16–17
plums 104–07
pomegranates 144–47
popcorn 51
potassium 17
potatoes 40–43
potting mix 187
propagators 12–13, 187
pumpkins 64–67

quinoa 182–85

rainwater collecting 17
raised beds 15
raspberries 112–15
recycling 6–7
red cabbage 60–63
rhizomes 70, 86, 187
roots, adventitious 96, 186
rose hips 150

salad leaves *see also*
  microgreens
  beets 59
  carrot tops 56–59
  lettuce 24–27
satsumas 101
Savoy cabbages 63
seed starting mix 187

seed heads, for indoor flower arrangements 54
seedling trainers, toilet paper tubes 15
squash 64–67
sterile plants 6
stratification 178, 187
strawberries 96–99
succession planting 43
sunflowers 174–77
sunlight 10–11
sweet potatoes 44–49

tangerines 101
tayberries 115
teabags
  chamomile 150–53
  instead of crocks 17
toilet paper tubes, seedling trainers 15
tomatoes 28–31
turmeric 89

water 10–11, 17
watering cans 14
windowsills 10–11

Acknowledgments

The publishers would like to thank Tom Brown and West Dean Gardens for their generous assistance with photography. Thanks also go to Duncan Turner for design assistance, Alison Gardner for sourcing props, Ruth Ellis for indexing, Kathy Steer for proofreading, and Adam Brackenbury and Sunil Sharma for repro work and color correction.

Picture credits

The publisher would like to thank the following for their kind permission to reproduce photographs:
(Key: a-above; b-below/bottom; l-left; r-right; t-top)

Simon Akeroyd: 7tl, 191tl; **Alamy Stock Photo:** Anakumka 177tr, blickwinkel / McPHOTO / HRM 133, Connect Images / Bill Sykes 59tl, Karen Kaspar 155, Natalia Kokhanova 173tr, Westend61 GmbH 35tr, Maarten Zeehandelaar 117; **Dreamstime.com:** Chernetskaya 34bl, 34br, Gheorghe Mindru 129, Trong Nguyen 105, Olga Ovcharenko 141, Veronika Peskova 181tr, Zerbor 73tr, Jinfeng Zhang 163; **GAP Photos:** Elke Borkowski 65, Chris Burrows 57, Victoria Firmston 25, Nova Photo Graphik 171; **Getty Images / iStock:** Jonas Hanacek 33, ozgurdonmaz 145; **Jason Ingram:** 47, 111tr, 113, 151, 153bl, 161tl, 161tr, 181bl, 183; **naturepl.com:** Nigel Cattlin 49; **Shutterstock.com:** anomaly026 58bl, Brzostowska 97, Djakandakandaduo 125, Maciej Dubel 111tl, Sergey Dudikov 46tr, ffolas 91, Freelancerbird643 176br, Geshas 107tl, Angela Lock 58br, Mirza Tanjia Nasrin 169bl, Nonavia 139tr, Irina Shatilova 115tl, studiomiracle 123tl, Dmitry Surov 134bl, VelP 157tr.

Senior Editor Alastair Laing
Senior Designer Barbara Zuniga
Senior US Editor Megan Douglass
DTP and Design Coordinator Heather Blagden
Production Editor David Almond
Senior Production Controller Steph McConnell
Editorial Director Ruth O'Rourke
Art Director Maxine Pedliham

Photography Jason Ingram
Illustration cover by Peter Bull Art Studio, all others by Louise Evans

First American Edition, 2025
Published in the United States by DK Publishing, a division of Penguin Random House LLC
1745 Broadway, 20th Floor, New York, NY 10019

Copyright © 2025 Dorling Kindersley Limited
Text copyright © Simon Akeroyd 2025

Simon Akeroyd has asserted his right to be identified as the author of this work.

25 26 27 28 29 10 9 8 7 6 5 4 3
005–344849–Mar/2025

All rights reserved.
Without limiting the rights under the copyright reserved above, no part of this publication may be reproduced, stored in or introduced into a retrieval system, or transmitted, in any form, or by any means (electronic, mechanical, photocopying, recording, or otherwise), without the prior written permission of the copyright owner. Published in Great Britain by Dorling Kindersley Limited

No part of this publication may be used or reproduced in any manner for the purpose of training artificial intelligence technologies or systems.

A catalog record for this book is available from the Library of Congress.
ISBN: 978-0-5939-5995-4

Printed and bound in Canada

www.dk.com

This book was made with Forest Stewardship Council™ certified paper—one small step in DK's commitment to a sustainable future. Learn more at www.dk.com/uk/information/sustainability

About the author

Simon Akeroyd has written more than 30 gardening books. He shares gardening tips and advice on his social media pages, which currently have more than 1.5 million followers.

Simon runs one-day gardening experiences based on his books as Simon Akeroyd's Gardening Academy, which he hosts in Surrey and Devon.

He was previously Garden Manager for both the Royal Horticultural Society (RHS) and National Trust. Gardens he has managed include Wisley, Polesden Lacey, Harlow Carr, Sheffield Park, Coleton Fishacre, Agatha Christie's Greenway, and currently Painshill in Surrey.

In his spare time Simon loves canoeing on the Dart River or exploring Dartmoor (and usually getting lost). He's currently writing a collection of Murder Mystery short stories with a horticultural twist.

Instagram: @simonakeroydgardenwriter
TikTok: @simonakeroydgardener
www.simonakeroyd.co.uk